STRATEGIC CONT

TOOLS AND RESEARCH TECHNIQU

T0257676

Erica Jorgensen

NEW YORK   2023

"I'm using this book immediately. Its fresh examples and focused research techniques will help my team serve our customers better than ever."

—Torrey Podmajersky,
author, *Strategic Writing for UX*

"Every content pro wants to increase the influence and respect of their team. Using Erica's end-to-end strategy and tactics, they can finally win over those tough stakeholders by transforming subjective opinions about content into objective research insights."

—Jonathon Colman,
Content Design Practice Lead at HubSpot

"If you've been looking for ways to make sure that your content is customer-centered and has real business impact, this is a must-have field guide for content evaluation and testing."

—Sheila O'Hara,
Head of Content Design, Slack

"Reading *Strategic Content Design* is like having a personal mentor quickly teach you the most important aspects of content creation and content research. While it would be ideal to have a dedicated user researcher on our projects, it's extremely valuable for digital teams to know how to gather customer feedback. Many of the insights that Erica shares would take years of experimentation and on-the-job analysis to learn on your own. If you would like to skip ahead to better content, then read this book."

—Michelle Mazurek,
Senior Manager, User Experience at Amazon

"This is me, shouting from the rooftops: *Strategic Content Design* belongs in the hands of absolutely anyone who cares about content in UX—by which I mean *EVERYONE*. Erica Jorgensen shows you how to create actionable content standards, put together an effective research plan for better content outcomes, and share your work across your organization."

—Kristina Halvorson,
CEO and Founder of Brain Traffic

"If you're writing words for a user experience without understanding their impact on your audience, you might as well be writing in a language they don't understand! Erica Jorgensen's book gives you all the tools you need to understand your users, measure your writing's effectiveness, and create clarity and quality. Read this book and go forth to make your users' lives better through effective content!"

—Andy Welfle,
Head of Content Design, Adobe, co-author of *Writing Is Designing*

"Good content answers our questions. But good content comes from good research—and in *Strategic Content Design*, Erica Jorgensen brings an astute and unflinching eye to research and testing techniques to engage the people, planning, and processes that shape content. Ready to ask tough questions? Roll up your sleeves; this is a book that prepares you to do the work."

—Margot Bloomstein,
Brand and Content Strategist and author of *Trustworthy:
How the Smartest Brands Beat Cynicism and Bridge the Trust Gap*

"Erica's book helps you get actionable insights to write truly user-obsessed content for your digital thing. The guidance and practices she outlines to get those insights and craft useful, usable, even delightful, content will be an indispensable resource…the kind that sits on your desk always at-the-ready when you need it."

—Meghan Casey,
owner, Do Better Content Consulting and author of
*The Content Strategy Toolkit: Methods, Guidelines, and Templates
for Getting Content Right*

"Erica Jorgensen's guide to research-driven content is packed with ways to get the best out of everyone whose work touches content. It will help any content professional build a strong foundation, navigate past content's biggest obstacle—unwelcome opinions—and toward content as a team sport that helps a whole organization win."

—Jane Ruffino,
Content Design Consultant and UX Writing Course Director, Berghs
School of Communication, Stockholm, Sweden

Strategic Content Design
Tools and Research Techniques for Better UX
By Erica Jorgensen

Rosenfeld Media

125 Maiden Lane

New York, New York 10038

USA

On the Web: www.rosenfeldmedia.com

Please send errata to: errata@rosenfeldmedia.com

Publisher: Louis Rosenfeld

Managing Editor: Marta Justak

Interior Layout: Danielle Foster

Cover Design: Heads of State

Illustrator: Danielle Foster

Indexer: Marilyn Augst

Proofreader: Sue Boshers

ISBN: 1-959029-57-6

ISBN 13: 978-1-959029-57-1

LCCN: 2022951877

Printed and bound in the United States of America

This book is dedicated to content creators of all kinds—
user experience writers, content strategists,
content designers, and more. I wrote this book
to shine a spotlight on the critical role of content
in creating effective digital experiences and
bringing greater respect, understanding, and appreciation
to the immense effort that goes into our work.

HOW TO USE THIS BOOK

Who Should Read This Book?

Strategic Content Design: Tools and Research Techniques for Better UX is intended for content professionals of all types—strategists, designers, copywriters, managers, and operations managers—and for leaders of teams of these content professionals.

It's also useful if you're part of a user experience or product team and you work with content creators—if you're a product manager, visual designer, UX researcher, or software developer. Reading this book will impart a greater understanding about the thought and care required in choosing just the right words. It will also build an appreciation of the critical role of content in connecting with your audience and customers, which results in improved business performance.

What's in This Book?

In *Strategic Content Design: Tools and Research Techniques for Better UX*, you'll learn how content research can transform your content team—bringing new energy and enthusiasm for their work and gaining attention and respect from teams of all types across your company (product management, product design, user research, operations, and engineering). You'll also get a toolbox with hard-won methods, best practices, and proven tips for conducting quantitative and qualitative content-focused research and testing. Use this book to:

- Create a strong content research program that builds positive energy, inspires your team, and draws attention to the importance of content to your products and business.
- Determine which methods and tools are ideal for your team's content research needs.
- Identify which specific words or content elements to test. Efficiently analyze your research results to uncover valuable insights about your audience that will contribute to a measurably improved customer experience (and therefore improved business results).
- Show why content research is worth the time and effort.

- Elevate the role of content design in your company, proving that content is key to creating an outstanding customer experience—and to improving your company's bottom line by saving your company on costs and increasing revenue.

- Get buy-in and support from colleagues inside and outside of content, leading to improved relationships across teams.

What Comes with This Book?

This book's companion website (🐏https://rosenfeldmedia.com/books/strategic-content-design) contains a blog and additional content. The book's diagrams and other illustrations are available under a Creative Commons license (when possible) for you to download and include in your own presentations. You can find these on Flickr at www.flickr.com/photos/rosenfeldmedia/sets/.

FREQUENTLY ASKED QUESTIONS

Why is research for content design important?

Content research can be directly tied to or correlated with business impact. If you know confidently what words and phrases are preferable to your audience and use those words in your content, you're going to help your company decrease costs (by reducing calls to your customer service department). You're also going to strengthen the business by creating more revenue, making customers more loyal to your company (increasing customer "long-term value," or "lifetime value," or LTV). See Chapter 11, "Apply Insights and Share Business Results."

With this increase in business impact comes greater respect for the role of content creators in your company. That respect, in turn, can lead to many other wonderful things: more attention for your team; opportunities to present the results of your work in business review meetings; a deeper, broader understanding across your company about the importance of content; and better support and recognition for your team, such as more resources, and even increased staffing and promotions.

I'm interested in running content research studies, but how do I know what content to focus on for evaluation? It's hard to know where to start.

It's natural when starting your content research program to want to evaluate *all* your content. Instead, a wise first step is to think about what major launches you have coming up. Is there a new product launching, a new feature that's being released, or a big promotional campaign to win over new customers? If you have a few weeks of time before such an event, it's a great idea to evaluate the content that's related to it.

If you don't have a launch or campaign on the horizon, you need to think about what your company's Most Important Content, or "MIC," is. For further details about how to identify your MIC and understand what specific content is smartest to prioritize for content research studies, see Chapter 4, "Evaluate the State of Your Content," and Chapter 5, "Identify Your Content Research Goals."

What are a few essential steps to take prior to doing content research?

First, you may need to take some important steps like defining exactly what "good" or "great" content is, from your point of view as a content expert. Second, set yourself up for success by making sure that your content is in solid shape before running any content research studies on it. (See Chapter 3, "Identify Your Content Quality Principles" and Chapter 4.)

Once you're confident about what you and your team mean when you refer to content that's good or great, update the content you're going to evaluate so that it reflects your "good content" principles. Only after you've taken that step and feel confident about the quality of your content should you start the process of evaluating that content using content research.

What types of content research questions are there?

Content research takes many forms. There are endless ways to go about it, although there are two basic types of research: quantitative and qualitative. Quantitative research measures the "what": What percent of your customers prefer one word over another, or what number of customers out of a sample of them are likely to take the action you want them to? Qualitative research reflects the "why": Why do customers think or act the way they do? (See Chapter 1, "The Power of Content Research.")

Depending on what questions you want your research to answer (see Chapter 5) you may want to use one or more of the typical content research study categories: Actionability, Audience-Specific, Clarity, Comprehension, Hedonic (or Sentiment), Naming, and Preference studies. There are a few basic content research question formats that can be used for all of those research categories, including multiple-choice, rating-scale, and open-ended questions. (See Chapter 7, "Craft Your Content Research Questions.")

CONTENTS

FOREWORD

People come to your website or app for the content. Of course, the visual design must be appealing. The interaction design must give users pathways that are logical to them. The technology must be quickly responsive. But that's all to support the content that people need.

Menu items, buttons, and terms that Search finds are content. The words and images that answer users' questions and take them through their tasks are content. Metadata that helps your SEO is content.

And, as Erica Jorgensen writes on page 101 of this excellent book, "Content is a product that requires care and feeding."

Content must be both useful and usable to the people whom you want as users of your website or app. For decades, I've taught this definition of usability (also adopted by the U.S. government and other countries as the definition of plain language):

> Whatever you create is usable only if people can
>
> - Find what they need.
> - Understand what they find.
> - Use what they find to meet their needs.

In this book, Erica answers the major questions that definition brings us to:

- How can we know that our current content or our planned content is useful, usable, and also engaging?
- How can we know we are using words and images that resonate with the people we want to visit our website or use our app?

Erica includes techniques like content principles and style guides that you can use even before taking your questions to users. She gives you detailed ideas for doing "quick and clean," as well as more extensive research with users so you can find just the right words and images to fulfill your content strategy.

This book supports and empowers you as a content strategist, content designer, content writer, or UX writer to add research to your toolkit. But Erica also urges you to collaborate with others as you do this. And I urge you others (designers, UX and CX people, marketing specialists, and managers) to read this book and better understand the need to and how to be sure your content is working as well as it can.

Strategic Content Design complements and adds to the 🐘Rosenfeld Media library, especially to

- *From Solo to Scaled* by Natalie Marie Dunbar
- *Surveys That Work* by Caroline Jarrett
- *Writing Is Designing* by Michael Metts and Andy Welfle
- *Content Everywhere* by Sara Wachter-Boettcher
- *Managing Chaos* (about web governance) by Lisa Welchman

Enjoy and learn from Erica's examples and very practical advice, as well as from the other voices and their stories that populate these chapters.

—Janice (Ginny) Redish,
author of *Letting Go of the Words—
Writing Web Content That Works*

INTRODUCTION

The Magic of Content Research

Content research can be truly magical—*so* magical that it can change lives and businesses, as this story illustrates.

Back in 2019, my Microsoft colleague, Trudy, was working with a product manager and product designer on a tricky challenge: clarifying Microsoft's monthly customer invoices and helping customers understand them more easily. On the surface, the invoices looked simple enough. Each invoice included information about the software that the company was using, the number of people in the company who were licensed to use it, and the overall monthly charges, tax, and any account credits (see Figure I.1).

Microsoft									September 2019 Invoice Date: 09/23/2019 Invoice Number: Due Date: 10/23/2019 **122.91 USD**	
Office 365 Business Essentials										
Service Period	**Days**	**Qty**	**Annual Price**	**Charges**	**Discounts**	**Credits**	**SubTotal**	**Tax %**	**Tax**	**Total**
08/23/2019 - 07/22/2020	335	5	60.00	274.59	0.00	0.00	274.59	7.25 %	19.91	294.50
08/09/2019 - 08/22/2019	14	5	60.00	11.48	0.00	0.00	11.48	7.25 %	0.83	12.31
08/08/2019 - 08/08/2019	1	4	60.00	0.66	0.00	0.00	0.66	7.25 %	0.05	0.71
07/23/2019 - 07/22/2020	366	3	60.00	-180.00	0.00	0.00	-180.00	7.25 %	-13.05	-193.05
07/23/2019 - 08/07/2019	16	3	60.00	7.87	0.00	0.00	7.87	7.25 %	0.57	8.44
SubTotal				114.60	0.00	0.00	114.60		8.31	122.91
Grand Total				**114.60**	**0.00**	**0.00**	**114.60**		**8.31**	**122.91**

FIGURE I.1

This earlier design was very simple, but it left customers confused about how the total bill was calculated.

While the invoices *appeared* to be simple, they were not providing customers with enough clarity. Microsoft customers were receiving their monthly invoices and then calling the customer-service phone number, trying to understand just how their monthly total was calculated. With millions of customers around the world, "many customers" with questions translated into a huge operating expense.

It was Trudy's task to work with her designer and create a more helpful version of the invoice that would prevent customers from calling customer service. The result: The designer added more white space and created separate sections for "charges," "credits," and "taxes." The revised invoice certainly *looked* more informative and helpful. However, when it was "stress-tested" with customers in usability testing, it became apparent immediately that the initial makeover of the invoices didn't achieve its goal.

Content Research to the Rescue

At the time, Trudy and I worked on the same product content design team. She shared with me how frustrating it was to spend hours working with the product designer to change the invoice layout and structure, only to receive disappointing feedback. She was also getting some pushback from the product manager and designer about adding details to the invoice; however, no one wanted additional content to potentially bump a one-page invoice to two pages, or make the invoice appear more complicated.

I suggested to Trudy that she take advantage of content research—a fast, straightforward way to get feedback directly from customers to find out what they were thinking, what content was clear to them, and what details or context was missing, so we could create a stronger customer experience.

Trudy was open to this idea. The challenge, though, was that our team at the time didn't have many resources to conduct content research. There was a UX research team, which helped with the "stress testing" mentioned earlier. But, like the content design team, the UX researchers were overstretched. If Trudy wanted the UX research team to help with setting up additional research time with customers, she had to request that time be allotted in the UX team's list of monthly sprint projects several weeks ahead of time. (To be clear, the UX research team was more than willing to collaborate. They simply needed to prioritize their work and couldn't take on an additional workload without taking other projects off their plates.)

We didn't have the ability to wait. The product manager was under pressure to get this problem solved quickly.

Trudy and I reported to the content design team director, Sheila. Sheila had an idea: She knew there was a smaller team within our broader Microsoft 365 organization that had access to a UserTesting account. Perhaps they would be willing to let us use UserTesting for this project? That way, we'd be able to set up an online study, share it with a sample group of customers—or people very similar to our customers—and hear their feedback, effectively helping us know exactly what information customers needed to fully understand the invoices.

Sheila was correct—the team with the UserTesting account was willing to give us access to it. Terrific!

Trudy and I set up and ran a couple of quick, short content research studies that effectively pinpointed which *exact details* customers needed to understand their invoices. We ran a Clarity test and a Comprehension test (see Chapter 7, "Craft Your Content Research Questions").

Long story short, the "aha" moment from the UserTesting content research sessions was this: customers said they needed the mathematical formula for how the invoices were calculated to be included *on the invoices themselves*. While customers stated they were unlikely to actually bother to take the time to plug in the numbers from their invoice to double-check the accuracy of the invoice's total, simply *listing the formula on each invoice* was key to building customer confidence and trust. Customers provided feedback such as, "I'm not about to use a calculator and do the math myself, but I can for sure tell you that having the formula right there helps me know how the [invoice] total came to be. I like that a lot."

Here's a close-up of the formula: *Licenses in the billing period x The Monthly or Yearly Price per license/Days in the billing period*, also shown in Figure I.2.

Microsoft

September 2019
Invoice Date: 09/23/2019
Invoice Number:
Due Date: 10/23/2019

122.91 USD

Office 365

Previous payments
Your previous payments are applied to pay for charges during this billing period.

Service period	Details	Licenses in service period	Yearly price/license	Charges	Discounts	Credits	Subtotal	GST %	GST	Total
07/23/2019 - 07/22/2020 (366 days)	Subscription charges already paid	3	60.00	-180.00	0.00	0.00	-180.00	7.25 %	-13.05	-193.05
Subtotal				-180.00	0	0	-180.00		-13.05	-193.05

Changes during this billing period
Changes made to your subscription during this billing period show as additional charges, or funds returned to you. Charges are prorated for the number of days impacted during the billing period.

Formula for charges
Licenses in service period X Monthly (or Yearly) price/license X (Days in service period /Days in billing period) = Charge

Service period	Details	Licenses in service period	Yearly price/license	Charges	Discounts	Credits	Subtotal	GST %	GST	Total
07/23/2019 - 08/07/2019 (16 days)	Charges before changes to this subscription	3	60.00	7.87	0.00	0.00	7.87	7.25 %	0.57	8.44
08/08/2019 - 08/08/2019 (1 day)	License change (+1)	4	60.00	0.66	0.00	0.00	0.66	7.25 %	0.05	0.71
08/09/2019 - 08/22/2019 (14 days)	License change (+1)	5	60.00	11.48	0.00	0.00	11.48	7.25 %	0.83	12.31
Subtotal				20.01	0	0	20.01		1.45	21.46

New charges
These are your charges for the next billing period for your current number of licenses.

Service period	Details	Licenses in service period	Yearly price/license	Charges	Discounts	Credits	Subtotal	GST %	GST	Total
08/23/2019 - 07/22/2020 (335 days)	Subscription charges for the next billing period	5	60.00	274.59	0.00	0.00	274.59	7.25 %	19.91	294.50
Subtotal				274.59	0.00	0.00	274.59	7.25 %	19.91	294.50
Grand total				**114.60**	**0.00**	**0.00**	**114.60**		**8.31**	**122.91**

FIGURE I.2.
The transformed invoice included a bit of detail for each section, to provide context and clarity. The mathematical formula for the invoice calculation, critically, was also included to build trust and confidence in customers.

With the insights she gleaned from the research, Trudy grew confident that the volume of (expensive) phone calls and emails to customer service would shrink.

The result: A couple of hours devoted to content research saved the day! The new invoice content was immensely successful in helping customers understand how their charges were calculated and far fewer customers called customer service for help with their bill—to the tune of $2.08 million less in annual costs. (You can find out how to calculate similar cost savings for your company in Chapter 11, "Apply Insights and Share Business Results.")

It's fair to say that these results were pleasantly shocking to the product team. The product manager was thrilled. The results of the invoice project were shared in the monthly business review, and the senior leadership team also took notice. Consequently, product managers reached out to us to find out just how we worked this "magic."

Those results had an additional halo effect. The work captured the attention of colleagues throughout the organization, effectively putting the work of the content design team in the spotlight. The upshot of that spotlight was an enormous, palpable boost in teamwide respect for the practice of content and for the content designers on the team. Where content had been taken for granted or, worse, left out of projects or looped in at the last minute before, now our work emphasized the critical importance of content to the customer experience *and to the business's bottom line.*

Granted, your company may be smaller, so the impact of your content research may not be this dramatic. But saving money for your company, no matter its size, is important. Similarly, content research can help you boost the revenue that your company earns, by transforming your content to be as engaging and effective as possible. Saving money and making money leads to a successful business. You, too, have the power to effect this positive change by using the content research techniques outlined in this book.

Enjoy!

—Erica Jorgensen
Seattle, Washington
December 2022

CHAPTER 1

The Power of Content Research

Here's an astonishing example of the power of content research. While I was working for a major health insurance company, the digital experience team was called into an urgent meeting. We were selling health insurance policies during the national open-enrollment period. That meant we had only 10 weeks to promote and sell health insurance plans. Two weeks into that 10-week window, the senior director of digital experience had alarming news to share: sales were only at a fraction of what was expected. Senior executives at the company were sounding the alarms. We needed to make up for lost time and "fix" the customer experience, immediately. *But what was going on?* After months of feverish user research with prospective customers, we'd created a visually appealing, simplified customer experience—or so we thought.

Like all insurance companies, we were offering three "flavors" of insurance plans: Bronze, Silver, and Gold. As you'd expect, the Bronze plans were the least expensive (though still pricey!). Silver plans were in the middle. Gold plans were the most expensive, but provided the widest choice of doctors, clinics, and hospitals.

In this meeting of the digital experience team, we collectively hypothesized about what might be happening from a customer experience point of view. All of the health plans were expensive. With this being the first year of mandated health insurance coverage, customers were understandably reluctant to choose a plan, because they were being forced to do so. People who previously had no health insurance were being asked to pay hundreds of dollars a month. And health insurance is an emotionally charged topic, and one that's famously complicated—it's one of the industries that's least trusted by the public, even less so than used-car salespeople!

For customers who qualified, government-funded subsidies were available to reduce the monthly cost—but required that customers jump through some application hoops and provide a lot of paperwork to provide proof of their income.

The digital experience team—including product managers, experience designers, content strategists—collectively put our heads together. Could we simplify the subsidy sign-up process? Part of that experience was out of our control; customers who wanted to apply for a subsidy were directed to a government website with complicated terminology—that is, when the site wasn't crashing from a huge volume of visitors. But, perhaps we on the digital experience team could provide a better online glossary and step-by-step guidance for customers, to help ease them through that process?

One of the product managers chimed in: "The sales of the Silver plans were so low," she said, "that perhaps there was a code error." Was the HTML buggy? Was the "Buy a Silver plan" call-to-action (CTA) button on the home page even working (which would be a huge embarrassment for our team, as we had done quality-assurance checks prior to the campaign launch date).

What seemed like a good hypothesis was shot down. The CTA button was working. Could there be something else cooking with this customer experience?

Collectively, we decided there was some more user research to be done—and quickly. We needed to find out why people were buying gold and bronze plans but avoiding the Silver plans like the plague.

A few hours later, a simple SurveyMonkey survey was shared with a sample of prospective customers. What we uncovered with that survey was gob-smacking, and helped save the day for the sales campaign. People replied to the survey and said things like, "I can only afford a Bronze plan. I would like better coverage but can't afford the Gold plan. And the Silver plan, that is not for me, because I'm not over 65."

Say what?

A pattern quickly emerged from surveying just 20 customers. Silver plans were perceived to be "different." Customers thought they were Medicare plans and only intended for people 65 years old and older. Silver plan, Silver Sneakers, Silver Fox, Centrum Silver vitamins…the "branding" of silver was getting in the way of our health plan sales! This was probably further complicated by how the website home page (and radio ads, and ads on buses, and social media promotions, and other advertising) was wholly focused on selling Medicare plans for the other 42 weeks of the year.

Damn.

With about two minutes of work, the content strategy team added two simple sentences to the home page, just above the Buy a Silver Plan call-to-action button, which made all the difference: "Silver plans are Affordable Care Act plans that provide a medium level of coverage. If you are over 65, shop for Medicare plans." That Medicare link sent customers to the Medicare plan landing page.

What a difference some clarifying content can make. Once that content went live, it was as if a light switch had flipped. Silver plan sales took off within the hour. Within a few days, sales were reaching the levels that the executive team had forecasted. It was like we waved a magic wand.

It was fortunate that we focused on the content, instead of trying to simplify the subsidy sign-up process. It was a lesson in never taking for granted how your content is being perceived by your audience.

Put Content Research to Work for You

Simply put, content research is a trifecta of goodness. First, it's an incredibly powerful tool for you, as a content professional. It makes your words work better and it creates a groundswell of influence for you and your team. Second, your customers benefit when you use clearer, easier-to-understand language. And third, it provides a boost for your business, because customers are more likely to trust your company and be loyal to your products, services, and brand when content speaks to them in an engaging, and relatable way. You know the phrase, "You're speaking my language?" Content research uncovers information about which specific words and phrases are clear and understandable and makes people feel recognized because you're talking their talk.

NOTE RAMPING UP THE CONTENT TEAM

Content teams are often staffed at a fraction of the levels of other digital product colleagues, such as visual designers, product managers, data analysts, and software developers. This understaffing needs to change! By showing your peers the sheer power of content and getting them to talk about the insights uncovered by solid content research, content research can also support the business case to improve content staffing.

What Is Content Research?

Content research involves asking your customers or audience for focused feedback on your content—for example, what they like, what they don't like, and why—and then using that feedback to improve your content. Sometimes this might be called *content testing*, especially if you're asking customers which words or phrases they prefer ("preference testing"). In this book, I'll primarily refer to it as *content research* because it more fully encompasses what this practice involves—providing insights that are key to you as a content creator. As with usability research, the insights gleaned from content research are like golden nuggets that can translate into a deeper understanding of your customers and their needs, which can result in improved business performance.

PRO TIP YOUR AUDIENCE

Sometimes it's not practical or possible to conduct this research with your actual customers or audience. In this case, you can use a proxy audience of people who are as similar as possible to your specific users.

Content research helps you accomplish the following goals, all of which contribute to your doing a better job as a content creator:

- Understand which specific words, phrases, descriptions, and messaging resonate for specific audiences—and which leave them confused or lacking confidence.
- Uncover insights about why customers prefer the words they do.
- Make content as customer focused as possible.
- Eliminate jargon from the customer experience.
- Reduce customer service requests and save your company money.
- Enable your customers to do what they came to your website to do—but more quickly and easily.
- Validate your content writing style guide.
- Inform your content design component library.
- Use your voice-and-tone guidelines.
- Emphasize just how much excellent content matters.

Understand What Resonates with Your Audience

The "what" of content research involves figuring out which words and phrases are preferred more than others, and the degree to which they're preferred. This "what" can also be referred to as *quantitative information*. For example, what quantity or percentage of your audience prefer one word or phrase over another? What specific words are clear, precise, and work best for your specific group of readers?

On a scale of 1–9, where 1 is "not at all likely" and 9 is "extremely likely," how do people from your audience rate their likelihood to use your app, based on a sample of content you share with them?

A QUANTITATIVE RESEARCH EXAMPLE

There are many ways to do quantitative research. Here's a basic example: Let's say you're developing a new feature for your product. You and your team worked together to come up with five potential names. There's a lot of debate about the merits of each—meaning that your team is arguing, and tension is building. Joe from the marketing team has strong feelings about Potential Name 1; you, as a user experience expert, have a strong hunch that Potential Name 2 will resonate better with your audience. Sound familiar? So, how do you decide which one to use and get on with launching this feature?

Content research to the rescue! It will uncover which name is preferable. Ask your customers or audience which potential name is more (or most) appealing: Name 1, Name 2, Name 3, Name 4, or Name 5. (You can obtain this audience feedback in a number of ways: ask a sample of your audience by phone or email or take advantage of a platform specifically made for such research, such as UserZoom, Microsoft Forms, Qualtrics, Survey-Monkey, or dscout.)

So you ask 20 people which name they prefer. You discover that 14 of them prefer Name 4, with the other four splitting their preferences among Names 1, 2, 3, and 5. This information tells you and your team that you're onto something with Name 4, because most of the customers surveyed appear to prefer it. Content research is a litmus test that tells you which way the wind is blowing. Chances are that your content—and, by extension, your customer experience, your product, and, yes!, your business—will be more successful if you use Name 4 (and less so if you choose Names 1, 2, 3, or 5).

A quick general note about content research: You can conduct content research on brand-new content that has yet to see the light of day with your customers. You can also use it to update or improve content that's already customer-facing, meaning content that's currently "live" and being viewed by your customers. You can improve your already-live content by taking words that you assumed were clear or preferred by your audience—but were discovered through content research to be not-so-clear after all—and replace them with words that your research identified as better.

Uncover the "Why" with Qualitative Research

Step 1—distilling quantitative information (the "what")—is amazingly powerful, as it gives you insights about what your customers are thinking. But content research gets even better. You can also find out why customers prefer the words that they do. This "why" research is referred to as *qualitative research*. And this is where even bigger golden nuggets—more like gold bars—can be uncovered so that you can better comprehend your customers' and audience's thought processes, which directly empowers you to create stronger, more effective content.

Quantitative plus qualitative information paints a fuller picture of your audience so that you can understand your customers' point of view: What do they like (or not), what do they understand (or not), and why? It pulls the curtain back so that you know what's really going on in their minds.

In addition, you can learn which specific words, phrases, and messages are clear to your audience, and which are fuzzy and need more explanation or description? What exact details are missing, if anything, so that your audience understands your business, services, or products? What information is extraneous? What does your audience like, or dislike—and why?

For example, when I started working on Microsoft's content design team, I was creating content experiences that guided customers as they bought and set up Office (now called *Microsoft 365*). It can be a complex, intimidating process, especially for small business owners who are counting every penny they spend. For each person in a company who uses Office, you need one license to be assigned to them, before they can start using the software. I thought the word "license" felt formal and not as customer friendly as it might be. Therefore, I felt it didn't entirely align to the company brand guidelines, which stated writing should be "crisp and clear" and convey that the company is "ready to lend a helping hand."

Instead of "license," I thought, why not use "seat," which essentially means the same thing, but doesn't have the potentially negative connotations of the word license. To me, "license" brings to mind the waiting in line at the Department of Motor Vehicles, bureaucracy, and wasted time. If "license" could be replaced by a clearer, friendlier word, the customer experience would feel a bit easier and lighter.

Well, I learned a lot from this content research study. We found that most customers who were asked (using UserTesting) about the word "license" felt just fine about it, to the tune of 17 out of 20 people surveyed. However, when we asked people *why* they preferred one word over the other, we discovered something unexpected and intriguing, and this content insight led the team to immediately change the user experience for the better.

What we found was that while the word "license" was preferred over "seat" many times over, and that most people understood the purpose of a license (in terms of why it was needed in order to start using the software), many people revealed to us that *they didn't know how many licenses a company needed.* This information bubbled to the surface when we asked people to tell us why they preferred either "seat" or "license." Some people thought you needed just one license for your whole company. Some thought you needed one license for each laptop, desktop, phone, or tablet. The truth is that you need only one license for each person using the software.

The implications of this nugget of knowledge were far-flung. If a customer thought they needed only one license, and bought one but had 10 employees, they would find that 9 employees couldn't use Office. They would need to return to the website to buy more licenses, and they'd likely feel frustrated and annoyed. On the other hand, if a customer thought their company of 10 needed 20 licenses (one for each employee, assuming each employee had both a laptop and mobile phone), they would overpurchase, and probably think Microsoft products were too pricey. Potentially, they would be less likely to renew their Microsoft software subscription and instead would be at risk for dropping their plan and jumping to the competition.

To prevent this customer confusion—*confusion we did not know existed prior to content research*—we simply had to clarify the customer experience. We immediately added a brief line of content to the shopping flow: **One license is needed for each user.**

There are business implications from this research! The net result of this action was a bit hard to quantify precisely, as we hadn't been measuring how many calls about license problems were being made to the customer service team each day. But we did know this: It was an immense improvement.

Say, for the sake of example, 1,000 people bought Office every day. We found through content research that 25 percent of the people didn't understand how many licenses they needed. So we were helping 250 customers *each day* to be more successful and feel more confident as they went through the steps to buy and set up Office for their company. This confidence translated into happier customers—customers who were more likely to stick with Office instead of moving to a competitor, and more likely to buy other Microsoft products. And if you extrapolate 250 people multiplied by 365 days a year, that's thousands of people who were in a better place thanks to crystal-clear content. In other words, content research makes your customers happier, and therefore your business more successful!

DIFFERENTIATING QUALITATIVE AND QUANTITATIVE RESEARCH

Don't let the names "qualitative" and "quantitative" intimidate or confuse you!

The difference between the two is simply that *quantitative* research refers to numbers: the quantity, or percentage, of something. For example, how many people think the call to action on your app is clear (or not)? Quantitative research is often conducted using surveys, which allow overall totals—and percentages and ratios—to be calculated. Out of 10 people who were asked, how many think that the call to action is clear? These kinds of research results are easy for you and your teammates (and managers) to understand quickly.

Qualitative research, on the other hand, is research that describes the characteristics or qualities of what is going on—the "why." Qualitative research is often conducted by asking people to write down or speak their thoughts through interviews or observation.

Sometimes you'll hear the two terms abbreviated as "quant" or "qual" research. (This is jargon, though in this case, it does seem to make the terms sound less intimidating!)

These two types of research are especially powerful when combined. Start with asking your audience a quantitative or "what" question, which will have responses you can count or quantify. For example, "Do you think this call to action is clear?" Or, "Which of the words in this list of 5 do you prefer to use to describe Product XYZ?" Then follow up that quantitative question with one that's intended to gather qualitative, descriptive information. A simple way to accomplish this is with an open-ended question. Ask your audience to explain *in their own words* why they answered the first question the way they did. For example, "Why do you prefer the word you chose? Please explain a bit about why you prefer it. Feel free to also mention why you don't prefer the other choices. Provide as much information as you'd like."

One important clarification to make right off the bat: Qualitative research is not higher quality than quantitative research. This is a misconception that people who are new to conducting research sometimes believe. Although qualitative research can often take more time, it can be an amazing way to discover remarkable insights about your audience that you wouldn't know otherwise. Combining qualitative with quantitative research is a powerful way to glean insights. For example, first ask a quantitative question ("Which word in this list do you prefer, if any?") and follow it immediately by a qualitative question ("Please tell me why.").

Make Your Content Customer Focused

As a content professional, you need to make sure that you're using words that your audience clearly understands. But when you can't avoid using words that are tricky and hard to understand—such as if you create content for a very specialized field, like finance, law, healthcare, education, or software—you may want to include definitions of any complicated words in your content experience. You can lean on content research to make sure that the definitions of any complex terms are as clear as possible.

A couple of important facts here: According to research from Deloitte,[1] companies that prioritize the customer experience earn 60 percent more revenue than companies that do not. Hopefully, that is enough inspiration to make sure that the words you're using are clear to your audience, instead of merely assuming that they are. (This evidence might also support you in obtaining a budget for content research tools and justifying the time investment needed for content research.)

Another fact: The Nielsen Norman Group, a longtime leading user-experience consulting firm, found that even "experts" in their field—lawyers and other highly educated people with advanced degrees—prefer simple, easy-to-understand words instead of jargon. See their study called "Plain Language Is for Everyone, Even Experts."[2]

---

1   Deloitte, "Customer-Centricity: Embedding It in Your Organisation's DNA," 2014, www2.deloitte.com/content/dam/Deloitte/ie/Documents/Strategy/2014_customer_centricity_deloitte_ireland.pdf

2   Hora Loranger, "Plain Language Is for Everyone, Even Experts," Nielsen Norman Group, October 8, 2017, www.nngroup.com/articles/plain-language-experts/

Eliminate Jargon from the Customer Experience

By gathering input about your content that comes straight from the mouths of your customers, you'll help raise awareness with your product-team colleagues about something that should seem obvious, but is often not acknowledged: Your customers don't necessarily use the same language as you and your coworkers! You therefore need to bring the voice of your customer front and center into your working process. Ongoing interactions with your customers help you create content that's jargon-free and not infiltrated with confusing or completely unintelligible words, terms, and phrases. At their most innocuous, jargon means nothing to your customers; at its worst, it leaves them feeling unconfident, stupid, incompetent, and wary of interacting with your content in the future.

Reduce Customer Service Requests

By improving your content and creating a clearer customer experience, you can reduce the number of phone calls, text messages, and emails from confused customers to your customer service representatives (and chat bots). You'll also reduce the number of customers who decide their customer experience was so poor that they no longer want to do business with your company and therefore abandon your company for your competition. Reducing both the volume of customer-service calls and number of lost customers quickly translates into significant money saved for your business. In other words, strong content builds and keeps customer loyalty.

Help Your Customers

Customers rely on the content on your website, app, or other digital experience for very specific reasons: to share information with them, guide them as they accomplish a specific task, or buy something. In other words, your customers have a job they need to do. Logically enough, in the user-experience world, approaching design work from this perspective is referred to as the Jobs to Be Done approach. (See *The Jobs to Be Done Playbook.*[3])

For example, consider a parent who has a teenager. That teen just passed their driving test and has a brand-new license. (And that parent might have some new gray hairs!) That family may use a car

---

3   Jim Kalbach, *The Jobs to Be Done Playbook: Align Your Markets, Organizations, and Strategy Around Customer Needs* (Rosenfeld Media, 2020).

Do you know how much each call, email, or text message from customers costs your company? Often, your customer service team will know how much it costs your business to provide customer service support. Imagine if, on average, it takes each customer service representative 10 minutes to resolve a complicated customer issue over the phone. If the reps get 200 calls a day, that's 2,000 minutes of phone calls, or 33.3 hours total.

You therefore need five reps working an eight-hour day to cover those 33 hours. (Yes, that leaves a little leftover time, but don't forget lunch and breaks.) If each rep earns $16 an hour, that's a cost of $128 a day each, or $640 a week each. Multiply that by five for the five representatives, and that's a weekly cost of $3,200. With 52 weeks in the year, that's an astonishing $166,000 annually.

Another way to calculate this is by the average cost-per-customer service contact. Since each rep earns $16 an hour and requires an average of 10 minutes for each contact, that works out to 27 cents a minute, or $2.70 per call. If you can leverage content to reduce the number of calls by just 20 calls a day, that's $54 saved each day. To project the yearly savings, multiply $54 by 365 days, and you get a savings of $19,710.

If you invest in a few hours of content research to remedy customer confusion, you can immediately start showing the return on investment (ROI) from content and how it saves your company money. You may not be able to reduce customer-service calls by 100%, but certainly you should be able to measurably reduce them. (Perhaps the money saved on customer service could be funneled into additional staff for your content team.)

insurance app to add that teen to their car insurance policy. Some more examples: A customer who is starting the onerous task of filing their taxes needs to know which IRS forms to use and what financial records they'll need as they fill out those forms. A person with an elderly parent who was just diagnosed with dementia needs to understand the various services provided by different types of long-term care facilities.

As a content expert, if you can ensure that you're using the clearest, easiest-to-understand terms to assist your customers as they complete their Jobs to Be Done, the happier your customers will be, and the more likely they will be to return to your site or use your product. (Note that each of these scenarios is emotionally charged! By creating clear content, you'll be helping people feel less anxious and stressed.)

Sometimes what customers need to do is buy things. If you work in ecommerce, content research can translate into more products sold and more money earned for your business, because content research points you to the words that appeal most to your customers and which specific details about products are most important to them. (Ever buy a dishwasher that should fit perfectly in your kitchen, but the box wouldn't fit in your apartment-building elevator? Details matter.) This sort of content clarity can also translate into fewer customer returns, again saving your company on costs and building the case for staffing up your content team because clear content translates into a stronger bottom line for your business.

Use Content Research for Style Guides

Many companies have established reference guides called *writers' guidelines* (and often *user experience–specific content guidelines*) that identify dos and don'ts for content teams to follow. Following these guidelines creates a cohesive, consistent experience for your customers. For example, on your app splash screen, do you ask customers to *sign in* or *log in*? How exactly do you spell your product and feature names and describe what they do?

NOTE MIND THE MARKETING AND ADVERTISING LAWS

Some of your content writing guidelines may be influenced by your company's legal team and by marketing laws. For example, the Federal Trade Commission (FTC) prohibits companies from overpromising about what their products can or cannot do, or from describing them in misleading ways. These laws are why Kraft Macaroni & Cheese lists the macaroni in its name first and the cheese second, because the product contains much more macaroni than cheese. If your company is involved in specific industries—such as the financial industry or environmental or "green" products—or if you market products to children—you need to be aware of the FTC laws that are intended to protect consumers. You should also include details about these laws in your writing style guide, for all content creators to be aware of.

Some additional entries typically found in a style guide include an A-to-Z word list of important words and frequently misspelled or misused ones. You can also often find which words are "on brand," to support the content voice and tone that your company is aiming to achieve, and which words and phrases should be avoided for specific reasons—including branding, sensitivity, diversity and inclusion, and ease of translation into multiple languages.

As you might imagine, style guides can be voluminous. As language and culture change, your style guide needs to keep up. A quick example: Referring to a person or a product as "crazy"—even "crazy good"—is now considered by many people to be offensive and improper, as that description is insensitive to people with mental illness. Many writing style guides therefore recommend avoiding the word "crazy."

One of the better-known style guides that's available publicly is the Mailchimp Content Style Guide (see Figure 1.1).

FIGURE 1.1
Mailchimp's Content Style Guide is one of the first digital content style guides to be shared publicly and provides brilliant advice, especially on catering voice and tone to a variety of customer content experiences, such as newsletters, legal content, and so on.

You've learned how content research can be used to make sure that you create strong content, and how it can be used to improve content iteratively that's already live. Another powerful way to use content research is to use it to validate your content writing style guides or content design library. In other words, to the extent you can, take advantage of content research to make sure that the words in your content style guide are the words that your audiences have told you are preferable and clear during content research studies.

By "to the extent you can," it means that you certainly don't need to validate every single word in your style guide. That could take years! Also, content teams are often understaffed, so it would be nearly impossible to conduct research on all the words and phrases in your style guide. That said, it's worth checking the "big cheese" words— words you use frequently, words that appear front and center on your app or get used over and over on your website, social media posts, sales team content, and so on.

Each time you run a content research study and learn about which words and phrases are clear and helpful to your audience, note this in your writing style guide. For example, because the Microsoft team's research confirmed that "license" is a word that customers tend to prefer instead of "seat," we noted "license" in our team style guide as a preferred word, along with a link to the research report that documented the research-study questions, participant responses, and resulting insights.

When documenting in your writing style guide or content design library which word or phrases have been validated, it can help to add an icon—such as a green check mark—next to each, for easy reference. Note the date that the research and its report were completed. (Depending on the volume of research studies you and your content team conduct, you may want to create a repository of searchable research reports.)

Validate Your Content Design Library Components

If you work in content design, you can use content research to validate your content design library components. Several companies like Apple and Atlassian have published their content design libraries online.[4] How do you know what language to use in each element

---

4   Atlassian content design component library, https://atlassian.design/content and https://atlassian.design/components

of your customer experience, including call-to-action buttons, error messages, empty states, and more? You can validate your team's language choices using content research.

Use Content Research for Voice and Tone

Because content testing is a conduit for hearing directly from your customers, it's also an excellent tool for developing or honing voice-and-tone guidelines. Voice-and-tone guidelines are often included in writing style guides and content design libraries as a special section.

Voice can be thought of as the unique stylistic qualities that identify your content. Voice often echoes the main characteristics of your company's brand. Is it playful and fun (like Taco Bell), inclusive (like Sephora), inviting and warm (like Starbucks), or sophisticated and refined (like Lexus)? (See Figures 1.2–1.4 for examples from these companies.)

FIGURE 1.2

The details page for the Lexus RX Hybrid automobile. The description of the car is heavy on words that convey a sense of luxury and richness. Words like "uncompromising," "exhilaration," "advanced," and "astonishing" combine to convey Lexus's luxurious tone.

FIGURE 1.3

The Taco Bell home page uses words like "bold," "epic," and "cravings" to convey a sense of fun and communicate the playfulness of its menu.

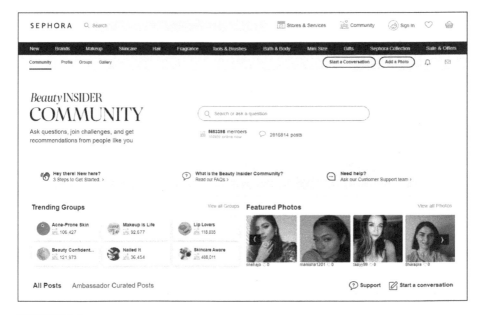

FIGURE 1.4

The Sephora Beauty Insider Community page includes content like "Hey there!" "Ambassador," and "people like you" to create a sense of belonging for its customers.

The tone of your content should align to the voice principles, although tone varies based on the situation. For example, if you're writing content that's used to acknowledge new customers, you may want to reflect an upbeat, optimistic, welcoming tone. If you're writing error-message content (for example, when a customer is providing their credit card number and they accidentally skip or mistype a numeral), you would write content with a straightforward tone, to help the customer fix the mistake and immediately get back on track.

One way to think about voice and tone is this analogy: *voice* is like your city's climate. Most of the time, it's predictable and falls within an expected range of temperature and precipitation. *Tone* is like the day's weather; it changes based on a variety of factors but doesn't typically veer outside of what you would expect for your climate.

Just as you can validate your writing style guide using content research, you can validate voice-and-tone guidelines using content research. Here's an interesting question to ponder: How did your company's voice-and-tone guidelines get developed in the first place? With all due respect to your colleagues, who says the words in your style guide or design library are the preferred and very best words to use, and how can they be so sure? You can turn to content testing to find out how well word choice and messaging approaches resonate— or don't—with your audience. It's an ideal way to collaborate with your marketing or brand colleagues to ensure that voice-and-tone guidelines support the creation of content that's appealing to your audience, and therefore is more likely to attract and keep them as customers. Validating voice-and-tone guidelines is also an excellent way to ensure that your own biases and those of your teammates don't get in the way of the creation of effective content. (Note once again: The words you personally prefer aren't necessarily the same words that your customers like best.)

Table 1.1 shows variations in voice and tone.

Brand: Starbucks

Voice description: Passionate, friendly, and warm

Tone: Varies, depending on the product, the content format and the customer experience or use case

TABLE 1.1 VARIATIONS IN BRAND TONE

Content for a Broad Audience: Website Home Page Content	Product-Specific Content: Starbucks Reserve (High-end) Coffee Packaging
Tone: Welcoming, almost neighborly	**Tone:** Inviting, inspiring, intriguing
Example: Whether you're searching for something new to warm your mug, seeking the best brew method for your favorite blend, or exploring our rarest offerings, you've come to the right place.	**Example:** EXPLORE OUR COFFEE Everything we do is in the name of coffee. This is the culmination of it all—our pursuit of the highest state of coffee experience, our relentless and ongoing innovation—all captured in your cup.
Comments: This content needs to speak to both those customers who prefer the basic drip coffees that Starbucks sells and those who like the (very) expensive small-batch espresso beans. It manages to sound welcoming to both.	**Comments:** This content needs to work hard to sell these very expensive coffee beans. Words like "culmination," "pursuit," "relentless," and "capture" convey an extreme appreciation for both the beans and the customer experience. With the last two words, "your cup," it's clear that Starbucks is extremely enthusiastic about these rare beans—the "passion" of the brand voice is intensified here into something close to obsession. It's obvious that Starbucks wants the reader to enjoy their coffee intensely.

Show That Excellent Content Matters!

Word choices can make or break a customer experience. Words can be the deciding factor between a happy, confident customer, and one who drops your product or company like a hot potato, abandons your brand to run to the competition, and then encourages their friends and family to do the same.

How to Conduct Content Research

Content research can often be conducted online using platforms like UserZoom or dscout. It can also be done in person (when that's safe to do).

As a content pro, you can run content research yourself. If you have a user research team, you can choose to collaborate with them to do content research or see if they can run content research on your behalf. (However, user researchers are often in great demand and frequently, unfortunately, understaffed.) If you have the budget for it, you can hire a separate third-party team to do it for you. Often, it's easier and faster as a content lead to run the content research yourself, because you thoroughly understand your company's writing guidelines, brand guidelines, and audience needs, and are therefore able to quickly plan and run studies yourself that will yield clear actions to take.

Each of these research approaches has its own pros and cons, which will be detailed throughout the book.

The great thing about content research is that, in many cases, it can be completed pretty quickly, so long as you scope your work to cover a manageable number of questions and focus on a select few, precise things you want to learn. If your audience is very specific or otherwise unusual, it can take more time to find the customers to participate in your research. That said, if you have a broad audience, you can often run a content research study using a platform like UserZoom and get helpful feedback and content insights in a matter of hours.

What to Expect from Content Research

Your content team will feel more empowered—and more connected to your customers—when you share content research among the team. When members of your feature team—product managers, visual designers, software developers, marketing managers, data analysts, and others—hear about the insights that your content team is uncovering using content research, you'll notice something magical starts to happen: You and your content team will start being treated with more respect. Respect will manifest itself in several ways: You'll be less likely to be left off the meeting invitations for a project kickoff and less likely to be skipped during the product-development process. You'll also be less likely to be looped in at the last minute to "clean up" content immediately before a project launch (the bane of content pros everywhere!). You'll also be invited to present at monthly business reviews and all-hands meetings. With content research, the role of the content creators and the importance of the words in your customer experience move to the spotlight. Instead of being the caboose in the product or digital experience development process, content moves to the front of the train and leads the way.

> **NOTE** **CONTENT IS NOT A QUICK, LAST-MINUTE STEP**
>
> Content should never be the caboose! If and when content creation is perceived by other teams to function as a quick last step before a project launches—a clean-up crew that merely proofreads—that's an ominous sign that content is underappreciated and undervalued in your organization. It shouldn't be this way, but often a content team needs to help educate those in other roles to understand the expertise and care required to create effective, clear content and how it translates to business impact. Content research can play a powerful, eye-opening role in this education.

WINNING OVER THE PRODUCT MANAGER

One of the most validating moments I had recently was when a product manager said during a sprint planning meeting, "We'll need to do content research for this project. We really need to make this customer experience super clear." In response, I first pumped my fist, and then told him I agreed wholeheartedly. When your product team understands the importance of content research, you know you've won them over to customer-informed content creation. When you have buy-in from your whole team to conduct content research, you'll be given the time and space to run your studies and analyze the research, share the content insights you discover, and create strong, effective, powerful content that reflects what you learned through research.

Be Diplomatic When Improving Content

A bit of warning when you conduct content research: You need to be careful when you evaluate content that's already "live" and out there in the world in front of your customers (sometimes referred to as *customer-facing content*). This is for two main reasons: one, the most obvious element here, is that you're criticizing someone's earlier work. Someone, or possibly many people, worked hard on that initial version of content. (Maybe it was you, if you are a content team of one or work on a small but mighty content team!) Whoever created that content did their best with the information they had to create the content as it currently stands. When you do content research on content that's already customer-facing, please do so gently and kindly, with a growth mindset—a mentality of let's always be learning and improving—instead of a critical, harsh approach.

The second thing to consider as you run content research studies is a managing-up issue. When content that's currently customer-facing is deemed lacking in some way (such as if content research reveals that significantly different terminology is preferable to what's currently being served up to your audience), you need to be careful as to how this is framed to your higher-ups. Again, try to take a positive, "let's-continually-be-improving" approach. The Japanese call this improvement-focused approach *kaizen*. You'll be better off taking a sensitive, emotionally intelligent approach than if you use a harsher attitude of "Wow, this current content is no good, and the research proved exactly that." Your leadership team will be more likely to embrace and support the practice of content research if you take a more thoughtful, *kaizen*-like approach. Proceed with tact and be diplomatic. (By the way, if you're ever presenting content research results to higher-ups and they ask why you didn't conduct research on *all* your customer-facing content, that's a perfect opportunity to ask for additional content staffing.)

What Content Research Is *Not*

While content research is sometimes a truly magic bullet for the content development process, there are a few things it's definitely not. Keep the following tips in mind as you take on content research, share content insights with your colleagues, and iterate and improve content to align it to the insights you discover.

It's *Not* a Crutch

Content research should not be relied upon to an extreme degree. Sometimes teams can get caught up in research and want to use it to answer every single content question every single day. While content research absolutely provides valuable insights and helps you make informed content decisions, it's not the be-all, end-all. When teams rely on research to an extreme degree, they often have difficulty using their own good judgment to make content decisions. Don't let this happen to you!

While it's terrific to be able to run content research on the words that are most important and essential to your customer experience (like, say, the splash screen of your new app, or your customer onboarding steps), it's impractical to think you can research every single bit of content you launch, or research every feature name, or every verb in every call-to-action button. The truth is, you don't need to use research for every content decision, nor should you. You must choose what content to spend your all-too-valuable time on, and that really ought to be the content that you know has the biggest impact on your customer experience.

Remember, there's no substitute for your experience as a content designer, content strategist, or content writer. You are a magical word person, and you therefore have an enormous amount of content knowledge in your brain and solid instincts from years of experience. Don't let content research erode your confidence!

It's *Not* A/B Experimentation

Content research is not A/B experimentation, and that's a good thing. A/B experimentation is the process of creating alternate versions of live content and displaying one version (the "control," or Version A) to one portion of your audience, and an "experimental" version (or Version B) of content to another portion of your audience. A/B experimentation has its time and place. It can be helpful, especially in ecommerce, to identify the words that are most appealing to your audience—the content that gets the most engagement or clicks. Sometimes A/B experiments are expanded to A/B/C/ or A/B/C/D/E experiments, also called *multivariate experiments*.

(A/B experiments are typically run until a "statistically significant" difference in engagement is seen. In other words, you need to be comfortable with and well-versed in statistical best practices to run them successfully. A/B experiments are therefore best suited to websites and apps with substantial daily traffic volume, so statistical significance can be reached after a limited period—days or weeks, instead of months.)

However, A/B experimentation is complicated and can be risky. It often requires assistance from your engineering team and your analytics colleagues to set up and run experiments and analyze their results. In addition, you're putting an untested version of content out into the world for your customers to see. If customers have a strong negative reaction to your experimental content, you'll be hearing from them!

Instead of managing complex, risky, and time- and resource-intensive A/B experimentation, you can use content research prior to A/B experiments to improve their effectiveness. That is, you can leverage content research as a first step, to identify appealing, engaging examples of content and run those freshly identified content examples against the "control" version. For example, say you have a call-to-action button label that's not performing as strongly as you'd like. You then think up five potential options for replacing the text. You can first run a content research study to discover which one or two out of those five options is most appealing to your target audience. Then you can set up your A/B or multivariate experiment with those "cream-of-the-crop" contenders, instead of merely guessing which may work most effectively. (If your content research study reveals a runaway preference, you can save time and skip the need for A/B experimentation altogether.)

It's *Not* "Regular" Usability Research

Content research can be conducted in collaboration with your user research team. But it's special in that it can often be completed by content pros *without* any help from the user research team.

(This is not at all meant to minimize the experience and wisdom of user researchers. They are irreplaceable and a key part of a successful product team. For example, I would rather not set up and analyze a tree test to identify the optimal navigation of my website. Similarly, I am in awe of the user researchers on my team and their ability to conduct moderated interviews, think on their toes, and adapt questions as they go.)

In other words, content research is something that content designers and content strategists can and ought to have responsibility for. In this respect, it's something that as a content pro, you can become known for in your company or organization, and it will further build your reputation as a content expert and customer-centric product team member who influences business impact.

Making It All Work for You

If you take away only one thing from this chapter, let it be this: *Content research builds the influence of and respect for your content team.* Content research can pave the way for so many things that come along with that influence and respect: being invited to project kick-offs, instead of being asked at the last minute to "wordsmith" content prior to a project launch; getting a seat at the table to represent content as an integral element of your company's digital experience; and the satisfaction of being recognized as a content professional with broad and deep specialized skills and expertise. To be clear: You shouldn't need content research to be able to prove your worth as a product professional. But until CEOs and VPs clearly understand the critical role of content in creating an exceptional customer experience, content research can play a pivotal role in transforming your work life.

As a content professional, you'll often hear unsolicited comments from your product teammates about how they can write, too. But not just anyone can create high-performing, customer-friendly, effective, clear, amazing content. In the same vein, anyone can use TurboTax to complete a simple 1040 tax return. But not everyone is a certified public accountant for a Fortune 500 company and responsible for profit-and-loss reports and complicated quarterly earnings statements. You are like that CPA, with expertise well beyond the average TurboTax user. The difference is like night and day. You deserve to get respect for your skills and expertise and their significant impact on the customer experience and your business's success.

When your company embraces content research, you'll elevate the levels of influence and respect that you and your content team receive.

CHAPTER 2

Leverage the Power of Clear Content

Clear language—sometimes referred to as *plain language*—is essential to understand before embarking on any content research. There are several ways that plain language comes into play for research. First, you need to use clear language in your content research study questions to help your participants easily understand them. Using clear language in your content research ensures that the input you receive from participants is as accurate as possible, so you can feel confident about using it to improve your content.

Second, when you use multiple-choice research questions in your content research studies, it's essential to present plain-language options to your participants. That way, you'll be helping your content research promote clarity in the finished content. For example, if you ask customers which of three potential names they prefer for a new coconut-macadamia cookie your company is introducing, and you include some long options that aren't as clear as they could be, you're not setting yourself up for success. Be careful what words you use in your research studies! You need to avoid presenting your test participants with fuzzy or overly complicated options that you wouldn't want to incorporate into your customer-facing content. Here's an egregiously dramatic fictitious example:

A. Mxyzptlk Cookies

B. Supercalifragiliciousexpialidocious Cookies

C. Kauai Coconut Crunchers

Avoid including in the multiple-choice options any names that you wouldn't or shouldn't publish—such as options A and B in the above example. The choices you list for your research participants ought to be succinct and clear, and aligned with your content voice, tone, and branding guidelines. That way, you can be sure that you're identifying and using the simplest, most effective words, phrases, and content for your audience, and you won't need to go back to the drawing board and repeat any research. Using short and clear options is also beneficial for your social media and advertising teams. (It's tough to squeeze a very long product name or any wordy content into a social media promotion or online advertisement, where space is at a premium!)

What Plain Language Is *Not*

To clear a common misconception right away: Using plain language does *not* mean "dumbing down" your content by using simplistic language. Rather, it merely means opting for easily understandable, clear language for the sake of your audience. As Sarah Winters of Content Design London said, using plain language is "opening up" your content to be embraced, used, understood, and valued (see Figure 2.1).[1]

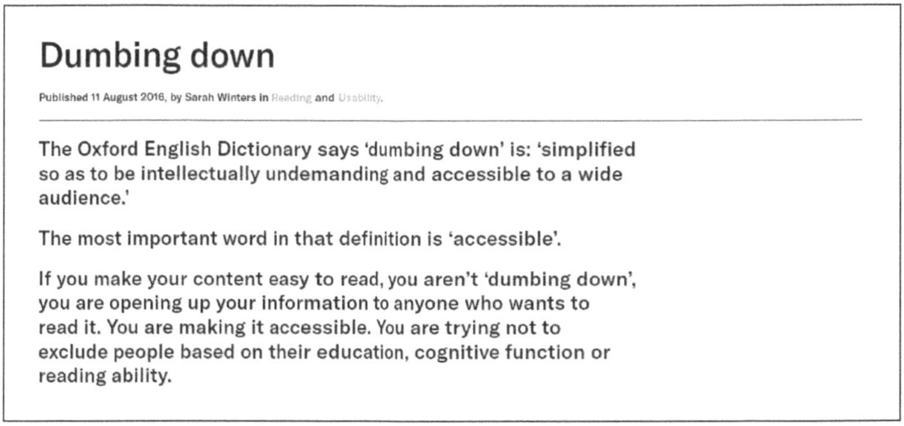

Dumbing down

Published 11 August 2016, by Sarah Winters in Reading and Usability.

The Oxford English Dictionary says 'dumbing down' is: 'simplified so as to be intellectually undemanding and accessible to a wide audience.'

The most important word in that definition is 'accessible'.

If you make your content easy to read, you aren't 'dumbing down', you are opening up your information to anyone who wants to read it. You are making it accessible. You are trying not to exclude people based on their education, cognitive function or reading ability.

FIGURE 2.1

Sarah Winters of Content Design London gives an explanation of why using simple language is important for accessibility.

Plain Language Means No Fluff

Plain language is language that's clear and efficient without any fluff. Content fluff is content that's loaded with extraneous information, sentences that meander, and lots of dependent clauses and prepositional phrases. It's content with wordy construction that's screaming to be tightened up and clarified. Content fluff can also refer to flowery, nonspecific content with long, complicated words that are a sign of a writer who is showing off, rather than working hard to help the audience.

Content fluff is everywhere, and it hinders your audience's ability to do what they need to do. Your audience is relying on your content to help them accomplish very specific things. Content fluff will at least

---

1   Sarah Winters, "Dumbing Down," Content Design London, https://contentdesignlondon/reading/dumbing-down/

Once, I had a software engineer disparage my intent to revise a user experience by replacing some complex content with plain language. He said to me (in front of many colleagues), "C'mon, we all went to college. Why would we want to create content for people who read as well as a 12-year-old? We're more educated than that."

Ouch. This statement stopped me in my tracks and flagged to me that I needed to provide some education to this engineer and the entire team about the value of plain language.

For starters, it's not true that all our colleagues went to college. And not all the company's customers are college-educated, given that we create products for people all over the world, for all sorts of purposes.

That said, I could appreciate this engineer's point of view and why he made that comment. Our team was creating content specifically for information technology professionals, or "IT pros" as our team called them, who were *typically* quite familiar with software terminology. However, just because we were creating content for a technically savvy audience, that did not mean we should have free rein to weigh down the content with hard-to-understand words and acronyms, when clear words and terms could be used just as easily.

Another consideration that I raised was this: not everyone in our audience is highly experienced. Some IT professionals might be newly hired, have switched careers, or otherwise not be as thoroughly and comfortably steeped in the lingo (or jargon) that's used by people who have worked in the field for years. The phrase, "You are not your audience!" applies here. It is a great mantra to repeat frequently when working in content.

slow them down and waste their precious time. At worst, content fluff will make your audience unsuccessful, and even prompt them to stop using your website or product out of frustration and seek a clearer, easier-to-use alternative.

Figures 2.2 and 2.3 show a couple of great examples of clear, tight messaging from Intuit—the company behind TurboTax—juxtaposed with longer, clunkier versions. These examples both come from the "Writing Small" section of Intuit's excellent content design guidelines.[2]

---

2   Intuit Content Design, "Writing Small," https://contentdesign.intuit.com/style/writing-small/

Let's get your W-2

To get started, we'll need a digital copy of your form—a PDF, JPG, or PNG.

OK, if you have a pdf copy of your W-2, we can use it to save time

When you upload it, we'll pull the info directly over.

Don't have a PDF? That's OK. We'll help you type your info.

FIGURE 2.2

Intuit highlights an example of clear, succinct writing and a wordier, clunky version of the same message. The first example is just 20 words long, while the longer version is 39—nearly twice as long!

Here's another example, also from Intuit's "Writing Small":

- Get expert help with Full Service

- You told us you wanted to hand your taxes off to a tax expert. Ready to continue?

FIGURE 2.3

Here is another example of short, tight writing from Intuit's content design guidelines ("Writing Small").

Check out these additional before-and-after examples for how to avoid fluff:

Before: Write sentences that are short and to the point.

After: Write short sentences.

Before: Be deliberately simple with your word choices.

After: Make each word count.

Before: Get rid of jargon.

After: Omit jargon.

Before: In the case where you simply can't avoid including a word that the average person on the street would struggle with, then define it.

After: If you need to include a word that the average person would find confusing, define it.

Certainly, if you work in a niche industry, your customer may not be "an average person." Maybe you work in the financial technology (fintech) field or for NASA. Terrific! But consider this: Have you ever been in a meeting where someone is dropping unfamiliar acronyms faster than Taylor Swift drops new records, and you really, really wish someone would dare to ask what they meant? And maybe, mercifully, a brave newly hired person or an outspoken summer intern dared to ask what the acronym stands for? And when the speaker defines the acronym, you hear deep sighs of relief from throughout the room (or Zoom call)? Be the brave person who identifies these acronyms and the words and terms that don't feel clear and clarify them for the sake of your customers.

Obliterating the fluff is sometimes difficult, but it's also exactly where content research can help out.

Figure 2.4 is an example of another website that's nailing simple language: Discord. The site has won a Webby Award, and it's well-deserved. The site clearly and effectively communicates the many different reasons why you'd want to use Discord, and how to get started.

FIGURE 2.4
The Discord home page clearly and concisely explains some different types of special interest groups that might consider setting up a Discord account—gaming groups, school clubs, or art communities. In very few words, it then explains why you'd want to set up an account with Discord: the app makes it easy to have frequent conversations.

Here's another example of fluff-free content, from the financial technology company Stripe, shown in Figure 2.5.

FIGURE 2.5

The Stripe home page exemplifies clear, tight writing, using just a few words to clearly communicate the app's purpose: to help companies of all sizes accept payments, make payments, and manage their finances.

DIFFICULT-TO-DEFINE WORDS

Here's an example of a hard-to-simplify word: "adjudicate." It's a term that posed a challenge to the digital experience team I worked with at a national health insurance company. It's tough to impossible to find a simpler substitute for "adjudicate" without losing some of its core meaning.

At first, the digital experience team tried to avoid the word whenever possible by reframing and recasting sentences, and writing around the word, bending over backward like Gumby to try not to include it in content, ever. That didn't work out so well. When we couldn't avoid using it, we decided to always provide a definition, because it isn't a well-known word (but it is an intimidating-sounding one!). We included the definition in a tooltip in digital communications like the website, or included the definition in parentheses (such as customer emails, or in printed communications like billing statements).

There are a million tips about how to write clearly. The tips and guidelines throughout this chapter are meant for you to keep in mind while conducting content research so that the content options you're sharing with research participants align with plain-language guidelines whenever possible. And note the choice of the word "guideline," and not "rule." Content rules can become tedious and can erode a content team's ability to be respected across feature teams. The connotation of "guidelines" is much more easygoing and implies flexibility and friendliness. Your content team will be perceived as easier to work with if you refer to *guidelines* instead of *hard-and-fast rules*. (This inflexible feeling seems to stem from the unpleasant experiences that many people had while learning grammar and writing "rules" way back in elementary school.)

THE ORIGINS OF THE EIGHTH-GRADE READING LEVEL GOAL

A popular rule of thumb for digital content that's been widely adopted by content teams over the past several years is this: aim for a sixth- to eighth-grade (U.S.) reading level for your content. This guideline has its roots in the 1940s. Melvin Lostutter, an assistant professor at Michigan State College—now Michigan State University—sought ways to help increase the circulation of newspapers. He found that newspapers were being written at a level that was five years above the average American's reading ability. Intriguingly, he also found that the articles that were easier to understand were more likely to be read.[3]

Lostutter was onto something, as 75 years later, plenty of other organizations, from the Content Marketing Institute to Content Design London and Nielsen Norman Group (in their much-cited and revised research report, "How People Read Online,"[4] are still recommending the use of straightforward, easy-to-read content for the sake of the reader. Only now, it's of even greater importance than in Lostutter's time, because it's more difficult to read and understand digital content compared to print.[5] Beyond that, writing simply and clearly is the right thing to do for inclusivity, accessibility, and sensitivity. For example, people who speak several languages and people who have ADHD or other cognitive challenges will be better able to understand your content when it's created with clarity in mind. Writing in a straightforward way is also beneficial for your translation or localization team, as it saves them time and effort.

---

3 Melvin Lostutter, "Some Critical Factors of Newspaper Readability," *Journalism Quarterly*, 24 (1947): 307–314, https://journals.sagepub.com/doi/abs/10.1177/107769904702400402

4 Kate Moran, "How People Read Online: New and Old Findings," Nielsen Norman Group, April 5, 2020, www.nngroup.com/articles/how-people-read-online/

5 Ferris Jabr, "The Reading Brain in the Digital Age: The Science of Paper Versus Screens," *Scientific American*, April 11, 2013, www.scientificamerican.com/article/reading-paper-screens/

Measuring Content Clarity

Content clarity can be determined by using a variety of tools, from apps like Readable[6] and Hemingway,[7] to the Editor tool built into Microsoft Word, to the Flesch-Kincaid formula (which ranks content on grade levels, as well as a readability scale that starts at 0 and works its way up to the rather unusual maximum value of 121.22).

These tools aren't wildly accurate when used to analyze short content, which of course includes user experience content. When given full sentences and paragraphs of text to analyze, these tools grow more accurate. That said, if you need to use these tools to analyze UX content or content designs—phrases or content that lacks subject/ predicate formatting—they can still be used as a helpful litmus test to provide feedback about whether your content is on the clear side, or if it's dense and hard to read.

The Flesch-Kincaid Formula

If you want to rabbit-hole on content clarity measurement, check out the Flesch-Kincaid readability entry in Wikipedia,[8] which gets into great detail about how the reading-ease formula was developed. Here's a sneak peek:

$$206.835 - 1.015 \left(\frac{\text{total words}}{\text{total sentences}} \right) - 84.6 \left(\frac{\text{total syllables}}{\text{total words}} \right)$$

The complex math formula for calculating Flesch-Kincaid readability or Reading Ease, which involves the average number of syllables per word and the number of words per sentence.

FIGURE 2.6
The formula for calculating Flesch-Kincaid Reading Ease is quite complicated!

---

6   Readable, https://readable.app

7   Hemingway Editor, https://hemingwayapp.com

8   "Flesch-Kincaid readability tests," Wikipedia, https://en.wikipedia.org/wiki/ Flesch%E2%80%93Kincaid_readability_tests

Here's how to find and set up the readability statistics in Microsoft Word. In Word, go to the File menu and then select Options (which appears at the very bottom, on the left side). There you will see the Proofing options shown in Figure 2.7. The annotations are mine.

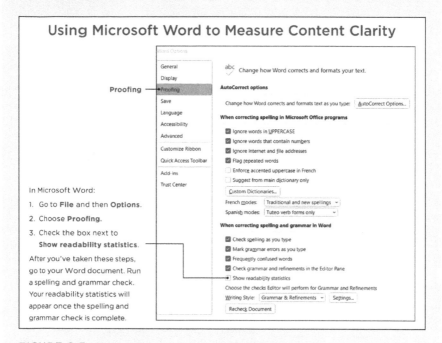

FIGURE 2.7

These steps outline how to find and turn on the Readability Statistics option in Microsoft Word.

To give you a feel for what content looks like at different reading levels, here are some samples of text with their rating levels:

From *The Cat in the Hat* by Dr. Seuss:

The sun did not shine.

It was too wet to play.

Flesch-Kincaid grade level = –1.5

Flesch-Kincaid Reading Ease = 116.4

This sample is so short and simple, it rates a *negative* grade level. Note that the words are one syllable each, which is a big factor in its rating.

Figure 2.8 shows another readability example, from the Expedia home page:

Save an average of 15% on thousands of hotels when you're signed in

Sign in

Sign up, it's free

List of favorites

Expedia Rewards

FIGURE 2.8
A promotion on the Expedia home page that's encouraging customers to sign in to their accounts before using the website.

From Expedia content:

Save an average of 15% on thousands of hotels when you're signed in.

The rating for this content is:

Flesch-Kincaid grade level = 6.7

Flesch-Kincaid Reading Ease = 69.9

Here's another example, also from Expedia:

Changes are coming to Expedia Rewards. We're introducing some changes to our program starting February 1.

Flesch-Kincaid grade level = 11.1

Flesch-Kincaid Reading Ease = 29.5.

As you can see, when you use words with multiple syllables (such as "introducing," "February," and "Expedia"), the readability drops dramatically.

One tip: If your company has a polysyllabic or unusual brand name, like "Expedia," scrub it out and recheck the readability score. If you do that for this example, you then get this:

Changes are coming to Rewards. We're introducing some changes to our program starting February 1.

The Flesch-Kincaid grade level falls to 9.3, and the Flesch-Kincaid Reading Ease increases to 41.3. (Keep in mind, lower grade level and higher Reading Ease equate to clearer content.)

Guidelines for Creating Plain-Language Content

It's easy to spot plain language when you see it. But how do you ensure that content is as clear and straightforward as possible? The following seven guidelines will help you write fluff-free content that is easier for your audience to understand:

1. Choose short words.
2. Use powerful, active verbs.
3. Use contractions.
4. Avoid jargon.
5. Avoid Latin.
6. Use clear words.
7. Write short sentences.

1. Choose Short Words—Three Syllables or Fewer

This three-syllable approach was recommended by Maggie Stanphill, the user experience director of Fitbit at Google. The guideline is pretty simple: If you spot a word that's three syllables long or more, try to hunt for a shorter replacement.

A key example of this is "ideate." Seriously! Why not just use "think up?"

Of course, there are times when you simply can't replace a polysyllabic word with a simpler one. But the three-syllable rule is generally a very helpful one. (It's also an easy one to teach to your product teammates in product design, product management, user research, and software development.) See Table 2.1 for examples of ways to simplify or replace longer words with shorter ones.

TABLE 2.1 SOME OVERLY LONG WORDS AND SIMPLER REPLACEMENTS

Overly Long Words	Tighter, Clearer Replacement Options
activate	start
additional	more
adhere to	follow
advise	tell
appreciate	value
associate with	connect
effective date	start date
eliminate	remove
enable	allow
ideate	brainstorm; consider; think
inform	tell
portal	dashboard
purpose-built	(Omit this description! People don't build things for no reason at all, do they? "Purpose-built" is awfully fluffy.)

PRO TIP HOW DO YOU KNOW IF A WORD IS COMMONLY USED?

There are many ways to find out if a word is commonly recognized or not. Check the online dictionary Merriam-Webster,[9] which has a handy gauge (for most words) that tells you how frequently people are looking up the word. If a word is searched for often (meaning, it's in the top 10% of words looked up on **Merriam-Webster.com**), then you know it's confusing to many people and likely warrants a replacement or an explanation. If you look up "ideate" on **Merriam-Webster.com**, you'll see it's in the top 4% of words that are searched for. That means not many people know how to define it, which means using it in your content will only confuse people.

---

9 Merriam-Webster, www.merriam-webster.com

2. Use Powerful, Active Verbs

Do your best to make the subject of your sentences do the action, instead of receiving it. In other words, keep your subject as the subject and not the object. When you use active verbs, your sentences stay short, while making the action much clearer and crisper.

Here's an example:

> **Jose tied his sneakers** is *active* construction. Notice how nice and short it is!

In contrast:

> **The sneakers were tied by Jose** is *passive* construction. It's also not as crisp as "Jose tied his sneakers."

<hr>

PRO TIP VERBS

Make each verb matter. Review your content and think hard about each verb and whether it's as clear as possible. Avoid linking verbs (like "be"). Use your new best friend, Visual Thesaurus,[10] to find the just-right verb for your needs. Also, when you can, avoid verbs that end in -ing (gerunds). When you omit gerunds, sentences feel much crisper and clearer. Gerunds can also be difficult to understand for people for whom English is not their first language.

3. Use Contractions

It seems like avoiding contractions is another writing "rule" that was cast down long ago by English teachers everywhere. This guidance now feels absolutely outdated, especially for digital content, as contractions both improve readability and create a more conversational, friendlier tone for your content.

---

10 Thinkmap Visual Thesaurus, www.visualthesaurus.com

There are only a few reasons to avoid contractions. One is if you're writing an amicus brief or some other kind of formal legal document. Another instance when it may help to avoid contractions is if you're writing warning or error messages for user experience (UX) content. In those cases, it can be helpful to avoid contractions, because doing so forces the reader to slow down a bit and focus on why the error happened and what exact steps they need to do next.

For example: "The app did not launch" can be more emphatic and clearer than "The app didn't launch." In this case, not is the most important word in the sentence. You're doing the reader a favor by writing out "did not" instead of condensing it into "didn't."

4. Avoid Jargon

An essential part of creating clear content is avoiding jargon whenever possible—and you'll find it's almost always possible, once you open your mind and consider just how much jargon might be lurking in your content. Jargon is something that sneaks into your on-the-job language without your realizing it, so keep in mind that what may be clear and understandable to you may be far from clear to your audience, especially new customers. If you think a term is jargon, chances are good that it probably is! Err on the side of simple language, and check Merriam-Webster if you're unsure.

A key example of jargon is the word "utilize." Why use it when the word "use" works just as well? (Actually, it works better, because it's shorter, and often in content design and UX writing, space is at a premium.) "Utilize" has creeped into vocabularies and sometimes content experiences because on the surface, it sounds more formal, official, or "smarter." But it's not, it's just longer! Opt for "use" instead, and you'll be better off. Your customers, your content's readability, and your translation team will all benefit if you choose the shorter, clearer option.

If you're at all in doubt about which words and phrases are clear and understandable to your audience and which need to be replaced or defined, run a content research study and let your customers or target audience tell you themselves.

Avoiding acronyms and abbreviations also falls under the category of jargon. In cases when you can't avoid an acronym, spell it out at first mention. (Keep in mind that thanks to the nature of digital experiences and the nonlinear way that content is consumed online, you may need to spell it out at first mention on your app, your website, your ads, social media posts, and so on.) For example, in the messaging platform Slack, the acronym "DM" stands for "direct message." If you're brand-new to using Slack, "DM" isn't necessarily clear. And if, as a content designer, you use unclear abbreviations, you're affecting your customers' confidence.

Here's the thing about jargon-y long words seemingly sounding smarter than simpler words (or, more "erudite," if you want to use a word that people seem to use when they want to sound like they went to Harvard). When customers are reading your content and encounter a word they are not familiar with, they may not go running to look up that word in Merriam-Webster. But what will happen is they'll feel unsure of themselves, even if just for a few fleeting moments. Perhaps they'll be transported in their minds back to a tough vocabulary test they took in sixth grade—when they also suffered the indignity of wearing braces with elastics—or maybe they'll flash back to the truly unpleasant experience of taking the Scholastic Aptitude Test (SAT) in high school.

Maybe your customer will feel a nagging twinge of insecurity. As a creator of digital content, you need to prevent this from happening. Work on building trust with your audience, instead of creating a chip on their shoulders. Using complicated, rarely used words, in an effort to sound "smart" or for any other reason, will only serve to undermine your relationship with your audience.

Figure 2.9 shows an example of business jargon, on the Intel home page. The language they use to describe the new Intel processor is muddy. What does "next-generation" mean? It seems that's up to you, the reader, to figure out.

FIGURE 2.9

The Intel home page is trying to generate interest in their newest core processor. What it fails to do is clarify why someone would want to get one. What kind of worlds would one want to create? Why are leaderboards mentioned? What does next-generation really mean, anyway? All of that is unclear. This content has a strong smell of "many cooks in the kitchen," where marketers, product managers, and other stakeholders may have weighed in too heavily during content creation and not let the content team do its job.

5. Avoid Latin

Who knows the difference between *e.g.* and *i.e.* off the top of their head? Many years ago, a colleague shared a mnemonic with our content team to help make these Latin abbreviations easier to discern. "E.g." is short for *exempli gratia*, which translates to "for example." Both *e.g.* and *example* start with the letter E. Ta-da! Easy to remember, huh?

Actually, no, not really!

Please write so that your readers don't need to think of mnemonic devices or go running for Merriam-Webster. Why not just say, "for example"? Straightforward content that's free of speedbumps wins every time. (For the curious, i.e. stands for *id est*, which is Latin for "that is.")

Also, Latin abbreviations are not helpful for individuals who are blind or have low vision and rely on screen readers. These abbreviations are also problematic for neurodiverse people as well (ADHD, dyslexia).

Knowing Latin *is* useful if you're a lawyer or a physician, or someone who enjoys tackling the *New York Times* crossword puzzle. But it has no place in customer-facing content, whether UX content, digital content like apps and websites, or content marketing. Latin is a type of jargon, often used for the sake of sounding "smart." For the sake of clarity, it's wiser to avoid it. Table 2.2 shows a list of common offenders and terms to replace them with.

TABLE 2.2 LATIN TERMS AND CLEARER REPLACEMENT OPTIONS

Word	Replacement
ad nauseum	tiresome
circa	about, approximately, or around
de facto	in fact
e.g.	for example
ergo	therefore
etc. or et cetera	and others
i.e.	that is
in lieu of	instead
ipso facto	by the very fact (Often, this is flabby content that can be omitted.)
per	each
per capita	for each; for each person
per se	by itself, or intrinsically
postmortem	after death
quid pro quo	this for that
sui generis	unique
verbatim	exact quote or repetition
vice versa*	turn around or change *Not only is this misused frequently, it's also often mispronounced!

(Note that some replacement suggestions are longer than the Latin term. This is OK! Clarity is more important than brevity.)

WHAT ABOUT MY LEGAL TEAM?

If you work in digital content, you likely collaborate with a lawyer or legal team who reviews your content to check that it's meeting privacy, security, and other legal standards. (If you don't, maybe you work for a startup that's pinching pennies and probably ought to find a legal team to work with.)

Your legal team may review content and return it to you with revision marks and paragraphs of "legalese" added, which can often include Latin terms and abbreviations. Or, worse, they may send you writing that's boldly marked "APPROVED BY THE LEGAL TEAM." They instruct you to use it as-is, no questions asked.

Please, please *do* ask questions. Not just to eradicate the Latin, but because as a content designer, you don't need to and indeed should not take such demands at face value. Always take the time to get to the bottom of things when lawyers request that you add legalese to your user experience. Keep in mind that while your company's lawyers are experts in the law, they are *not* experts in helping your customers understand things. That's your job, and it's damn important. Don't be afraid to collaborate closely with your lawyers and help them understand how your audience perceives content, and why unclear, intimidating-sounding words can erode your customers' trust.

Don't be afraid to ask: Why exactly is this legalese necessary? Find out what exact law or rationale is behind their request. Sometimes, lawyers ask that legalese be added not when it's truly necessary by law, but merely as a CYA ("cover your a**") move. This is similar to when people avoid using contractions in their writing, just because their eighth-grade English teacher told them to do so. Do some diplomatic, polite digging—what I call engaging in *Crime Scene Investigation: Content*. Find out if the legalese absolutely, positively needs to be added, or if your lawyer would merely prefer it be there, because they were trained to be overly cautious and risk-averse (and maybe, just maybe, are trying to justify their very high salary). You may need to document why you're pushing back and recommending clearer language be used instead of formal, complicated legal terms. That's okay—it's a step in the right direction for the sake of your customer experience.

A great resource for info on marketing laws is the Federal Trade Commission's Business Center,[11] seen in Figure 2.10, which includes details on digital advertising and other special categories that rightfully warrant deeper regulation: health claims, environmental marketing, and marketing to children.

Another helpful, searchable site with legal guidance is Cornell Law School's Legal Information Institute (LII).[12]

FIGURE 2.10

The Advertising and Marketing section of the Federal Trade Commission website. This includes helpful information for content strategists and content designers, including details about marketing laws for specific scenarios, including marketing to children and marketing that includes health claims.

---

11 Federal Trade Commission Business Center, www.ftc.gov/tips-advice/business-center/advertising-and-marketing

12 Cornell Law School Legal Information Institute, www.law.cornell.edu/

6. Use Words That Are Clear to Your Audience

This guideline is exactly where you can leverage the power of content research to guarantee that you are using words that your customers understand and are comfortable with. Here's a rule of thumb: *Don't assume anything about your audience. Work to find out what words, phrases, and content they understand, and what exactly they need help to understand clearly.*

The Nielsen Norman Group found that plain language is appreciated and *preferred* by audiences of all kinds, including lawyers, academics, and scientists with Ph.Ds.[13] Nielsen Norman's researcher, Hoa Loranger, summarized their findings: "Professionals want clear, concise information devoid of unnecessary jargon or complex terms. Plain language benefits both consumers and organizations." The difficulty is that distilling your writing so that it's plain and clear takes a lot of time and effort.

7. Write Short Sentences

There's a classic swashbuckler movie called *Zorro* (remade most recently in the 1990s) whose eponymous character wields a sword. When I see content that's long and complicated, I use the phrase, "Give that content the Zorro treatment." In other words, cut those sentences down the way Zorro would shred the shirts of his enemies.

It's helpful to avoid long, complex sentences for all types of content, but it is especially true for UX content, because your audience needs to scan for key information. Long sentences—those over 18 words or so—force your audience to slow down to understand, and that makes it tough for them to accomplish whatever Jobs to Be Done or tasks they're aiming to do.

---

13 Hoa Loranger, "Plain Language Is for Everyone, Even Experts," Nielsen Norman Group, October 8, 2017, www.nngroup.com/articles/plain-language-experts/

An easy way to give sentences a haircut: look for prepositions and work to reduce or omit them. The most commonly used prepositions in English are: *at, by, for, from, in, of, on,* and *with.* Content designers and content creators of all sorts rightfully hear sirens going off in their head when they see these words in content drafts (often written by product managers, product designers, or software developers who mean well, but who are not content experts). Here's a tip: Put parentheses around your prepositional phrases, to give yourself a visual cue. Then rewrite your sentences to omit them.

Weak, fuzzy, unnecessary content is often found lurking in prepositional phrases. Here are a few examples in Table 2.3 where prepositional phrases result in wordiness and impreciseness.

TABLE 2.3 PREPOSITIONAL PHRASE MAKEOVERS

Content Including Prepositional Phrase	Shorter, Tighter Option
at a distance	far
around 12:30 p.m.	12:32 p.m. (When omitting prepositions, use it as an opportunity to be specific.)
in addition to	also
in an effort to	to
in order to	to
out of an abundance of caution	carefully (This example has two prepositional phrases that can be edited!)
within a few minutes	soon

OMIT EXTRANEOUS PREPOSITIONAL PHRASES

Here's an example of a sentence makeover that's based on ridding the sentence of extraneous prepositional phrases.

Original version	By creating a shopping cart with multiple features with appeal to your customers, you will be strengthening your business for years to come.

This original version is 24 words long. It's wordy and not very convincing.

Step 1: Identify the prepositional phrases.

By creating a shopping cart (with multiple features) (with appeal to your customers), you will be strengthening your business (for years to come).

(Notice how the prepositional phrases take up a good chunk of the content length—about half of it!)

Step 2: Analyze each prepositional phrase.

"with multiple features": It's unclear what features you're talking about. It's a lost opportunity to not describe them. As a content designer, this is your chance to engage in some *CSI: Content* and obtain more details from your product manager, the brand team, or marketing—or anyone who can provide details.

For the sake of this exercise, let's pretend you asked the product team what features were most important, and they said that market research showed they were mostly related to shipping details. Add in that newly obtained detail, to transform the content from vague and fuzzy to concrete and clear.

"...with appeal to your customers": Well, this sounds awkward because it's the second "with" in the sentence. This is a prime example of how showing instead of telling is a more powerful content approach. How exactly do they appeal? This is dull.

"...for years to come." This feels like a wishy-washy Charlie Brown phrase. It doesn't add anything of value to the sentence. Trash it. And if your product team pushes you, saying that this is a super-important point that needs to be made, push back. Who says it's important? Did the customers say so? Just like the "multiple features" phrase above, do some investigation to find out what concrete information can be included to bring the content to life.

continues

Here's the transformed content:

After	Strengthen your business by creating a shopping cart with the shipping-option details your customers want.

13 words—that's nearly 50% shorter—and much more powerful sounding than the original!

This sentence still includes one prepositional phrase—"with shipping-option details." But it adds new information that's key to the reader, and it's a vast improvement over the original. Giving the Zorro treatment to two out of three prepositions isn't bad at all!

And here's the calculation, if you want to know exactly how much shorter the new, improved sentence is:

(New length – Original length)/Original length × 100

$= (13-24)/24 \times 100$

$= -11/24 \times 100 = -45.8\% = 45.8\%$ shorter!

It's stats like these that can make your feature teams stand up and take notice of the impact that your content team is making!

Great Resources for Writing Clear Content

For more advice on writing crisply, see Roy Peter Clark's *How to Write Short* and *Writing Tools: 55 Essential Strategies for Every Writer* (or really, anything written by Roy Peter Clark, as he's a word wizard extraordinaire with over 15 books to his credit).[14] More great references are *Nicely Said* by Nicole Fenton and Kate Kiefer Lee,[15] *The Reader's Brain: How Neuroscience Can Make You a Better Writer* by Yellowlees Douglas,[16]

---

14 Roy Peter Clark, https://roypeterclark.com

15 Nicole Fenton and Kate Kiefer Lee, *Nicely Said: Writing for the Web with Style and Purpose* (San Francisco: New Riders, 2014) and www.nicelysaid.co/

16 Yellowlees Douglas, *The Reader's Brain: How Neuroscience Can Make You a Better Writer* (Cambridge, England: Cambridge University Press, 2015).

and the classic *Letting Go of the Words: Writing Web Content That Works,* 2nd Ed. by Janice (Ginny) Redish.[17]

You can also read more about the outstanding guidelines from Content Design London[18] and the Plain English Campaign.[19]

Making It All Work for You

Here's the main takeaway about plain language and its relationship to content research: Time after time, you're going to find that your customers share that they prefer straightforward, simple, plain words and phrases over complicated ones.

And even if you work in a field or industry that has its own very specific vernacular, don't think for a minute that your audience knows what you're talking about when you use what your colleagues probably refer to as "commonly used industry terms."

If you were ever confused in a work meeting by a confusing, unfamiliar acronym and then breathed a sigh of relief when someone dared to ask what it stood for, you know this is a thing. If your colleagues balk by saying that a certain word or term ought to be fine to use in your content "because it's in the style guide," or because it's included in your design library, gently but firmly reply that the words and terms in your style guide *ought to be content tested* and not assumed to be clear.

Every word matters. As a content designer, you must refuse to tolerate fluffy and overly complicated words and phrases. *And you must fight back for the sake of your audience every time a product manager, designer, marketer, lawyer, software developer, or anyone else you work with suggests a fluffy or overly complicated word or phrase be used in your content.*

Plain content can help you achieve the goal of content design: to communicate clearly with your audience with minimal friction and minimal cognitive load, so that your customers can accomplish whatever it is they need to do.

---

17 Janice (Ginny) Redish, *Letting Go of the Words: Writing Web Content That Works,* 2nd ed. (Boston: Morgan Kaufman Publishers, 2014.) and redish.net

18 Content Design London, "Readability Guidelines," https://readabilityguidelines.co.uk

19 Plain English Campaign, plainenglish.co.uk

An Interview with Kathryn Brookshier,
Senior Researcher at Indeed

It can't be said more emphatically: Don't assume the words on your website or app are working well until you run research with actual customers.

Kathryn Brookshier, a senior researcher at Indeed—the world's leading job-search website—is an enthusiastic proponent of content research. "It's absolutely illuminating, the power of the words and what they're communicating to people. If the words aren't clear, you're not doing your business justice."

Kathryn says that at Indeed, much of the content research being conducted is focused plainly on comprehension. "We take a specific bit of content and ask participants basic questions about the content—information that's theoretically explained in the text. We ask what outstanding questions they have about the content, and often find we have been assuming things are clear that are absolutely not clear to them."

One example that had enormous business repercussions at Indeed was the recruiter/job seeker interview-scheduling experience. Job seekers attended hiring events and were able to directly sign up for interviews, selecting the date and time they preferred. Sounds simple! But what content research uncovered was that the content in both email and text messages was so unclear to job seekers that they didn't realize when interviews were confirmed. They therefore weren't showing up online for interviews—but recruiters were. That led to embarrassing experiences for the job seekers, who were trying to put their best foot forward as they actively sought a new job—and wasted time for recruiters, who expected eager job seekers to be ready and prepared to be interviewed.

After content research and subsequent clarification of the interview confirmations, the result (no surprise!) was increased attendance (and happier job seekers and recruiters).

Kathryn says another area where she and other Indeed coworkers use content research is updating navigation labels. Navigation tends to be additive, growing and growing more complex over months and years, often without a "forest-for-the-trees" evaluation of how it's currently working. In addition, navigation labels that worked a few years ago may not necessarily be as clear to customers as your business or products and features evolve.

"Research helps our content strategy team to be more easily able to do data-driven decision-making about their work," she says.

CHAPTER 3

Identify Your Content Quality Principles

Before starting your content research program, it's important to pause to assess the state of the content that you and your immediate content team have created. If your company's size makes it manageable and feasible, do a gut check on the state of content across your organization. If you work for a larger company, review and evaluate the specific body of content that your team is responsible for. What do you notice about the strengths and weaknesses of your content?

Taking this introspective step is practical, wise, and meant to prepare you for the work ahead, including important conversations that will be necessary to set the stage for a successful content research program. You need a sense of the overall quality of the content that's already been created before you work to take steps to improve it. You also need a solid feel for what qualities you and your team feel makes for great content, compared to content that's just okay, or plain not good at all. *What exactly are your content standards?* Who created them, and when? How are the content standards shared, inside and outside the content team? (And how aware of these standards is your content team, and people on your partner teams like product management, user research, and engineering?)

Having guardrails in place for content principles—and a standardized language around what high-quality, high-performing content looks like—will be helpful not only when your content team collaborates to make new content, but also when you go to partner with colleagues outside of content to develop content research studies. These content quality standards will also come into play *after* you run research studies, when you go to bat with your production or engineering team to make content updates and iterations that reflect the insights you uncovered with your research. Lastly, they'll be advantageous as you work to spread the word about your content research efforts, and can also serve to support you as you seek to build respect for and the influence of your content team across your team and company.

CONTENT HEURISTICS

Content heuristics are guiding principles that define what you and your team mean by "quality content." Often, content teams develop a short list of five or ten qualities they want their content to adhere to. (A couple examples: content needs a call to action; content needs to use inclusive language.) Heuristics can be very helpful when you have people from a variety of teams—such as product management and product design—involved in your product process. Frankly, heuristics can also help your content team stand its ground when people from outside your team share unsolicited opinions about the content.

The word "heuristic" is rooted in the Greek word for "discovery." What this means is that most of the time, your heuristics should provide terrific guidance as you create content drafts and also foster a sense of confidence about the content. (They can also help you create content more quickly.)

Perhaps the most famous set of user experience heuristics was created by the Nielsen Norman Group (https://nngroup.com) in 1994. (They are updated periodically.) Nielsen Norman's "10 Usability Heuristics for User Interface Design"[1]—as their name hints—don't mention content all that much. (Nielsen Norman Group is primarily focused on visual or product design.)

Here's one of the more content-focused heuristics from the Nielsen Norman Group's list of 10 heuristics: #4, Consistency and standards, as seen in Figure 3.1. In this heuristic, the Nielsen Norman Group recommends avoiding the use of multiple words that mean the same thing. ("Users should not have to wonder whether different words, situations, or actions mean the same thing.") What this means is that you should be careful to be consistent across all your user experiences. If you describe a specific feature as a "super widget" on your website, use that same description in your mobile app and all other customer-facing content—not an "awesome widget" or "amazing widget."

continues

---

1 Jakob Nielsen, "10 Usability Heuristics for User Interface Design," Nielsen Norman Group, www.nngroup.com/articles/ten-usability-heuristics

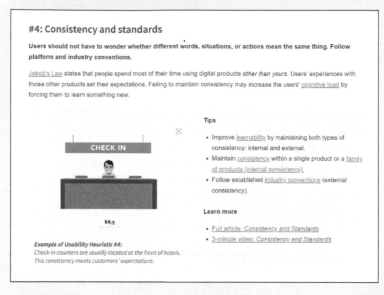

#4: Consistency and standards

Users should not have to wonder whether different words, situations, or actions mean the same thing. Follow platform and industry conventions.

Jakob's Law states that people spend most of their time using digital products *other than yours*. Users' experiences with those other products set their expectations. Failing to maintain consistency may increase the users' cognitive load by forcing them to learn something new.

CHECK IN

NN/g

Example of Usability Heuristic #4:
Check-in counters are usually located at the front of hotels.
This consistency meets customers' expectations.

Tips

- Improve learnability by maintaining both types of consistency: internal and external.
- Maintain consistency within a single product or a family of products (internal consistency).
- Follow established industry conventions (external consistency).

Learn more

- Full article: *Consistency and Standards*
- 3-minute video: *Consistency and Standards*

FIGURE 3.1

"Consistency and standards" is one of the Nielsen Norman Group's 10 usability heuristics. From a content perspective, being consistent means reducing cognitive load (or the mental gymnastics required to understand how to use a digital experience) by using the same terminology across all your content surfaces.

The Roles and Values of Content Principles

How does anyone judge the quality of content? If you've been in the business of content for only a brief time, you may hear others on your team (or on LinkedIn, social media, or at conferences) refer to content terms that are unfamiliar to you: conciseness, accessibility, or inclusive language. This section will help you better understand the many facets that can be combined to create excellent content. If you're an experienced content professional, you probably have a gut feel for what makes content successful: it reads smoothly, because words have been chosen carefully, and they are properly spelled and punctuated. It also follows guidelines for good grammar and clear syntax, so the reader's cognitive load is low, and there's no need to stop and re-read content to make sense of it. Sentences vary in length and structure, which keep the reader's interest.

In product content and content marketing, it's also essential that you include in your content a very clear next step or call to action (abbreviated as "CTA"; the plural is "calls to action"). You also need to ensure that your content reflects your company's voice-and-tone standards. (You can find more details later in this chapter, under "Content Voice-and-Tone Guidelines.") You also need to align your content to accessibility best practices, such as ensuring that your videos have captioning and transcripts, so they can be understood by people who cannot hear, as well as following inclusivity and sensitivity standards. For example, avoid addressing your audience as "you guys" when the audience includes people who identify as genders other than male. Inclusive writing also means not using insensitive, offensive terms like "master" when you mean "core" or "primary."

But do you have a name for all these content qualities, so that people on your content team—and across your company—have a common language to use when talking about content? How do you talk about content with people who don't create content day in and day out, and tend to stare at you like you're a Martian when you mention words like "syntax"?

These are all considered to be content standards, principles, or heuristics. Do you have yours documented and shared across your content team—and better yet, across the teams of stakeholders you work with frequently? Ideally, across your whole company? If your content team has a content Center of Excellence or Community of Practice, chances are that you do have content heuristics documented and in use. Lucky you! But if you don't, not to worry. You and your content team can create them. The tricky part will likely be taking a long laundry list of potential qualities and agreeing as a team on which principles are the most important. (See Chapter 12, "Build Momentum with Content Research," for more on Communities of Practice.)

A *Community of Practice* (*CoP* for short) is a morale-boosting support group. Many content Communities of Practice focus on how to raise the profile of the practice of content. If your company is small, you may want to create a cross-company content Community of Practice that welcomes all sorts of content creators: user experience/content design, content marketing, technical documentation, social media and external communications, internal communications, public relations, corporate marketing, and more. If your company is larger, it may make sense to have several coexisting Communities of Practice, each with its own specialty or focus.

Whatever your company size, make sure to cast a wide invitation and make the Community of Practice a welcoming, inclusive group. Inviting content professionals from outside companies to speak at your Community of Practice meetings is a great way to understand how other teams manage the challenges of working in content. For more pointers on starting and keeping a Community of Practice running, see the blog post in Figure 3.2, "How to Build Your Community of Practice," by Kristina Halvorson of Brain Traffic.[2]

Members of a Community of Practice meet regularly to align on messaging, share challenges, and suggest solutions. Is your content marketing team running a pilot with an artificial intelligence–powered platform that reviews content for them prior to publishing? That's good information to know. Some Communities of Practice also invite guest speakers from outside companies and organizations, to learn about what tools, practices, and processes are working for them.

---

2   Kristina Halvorson, "How to Build Your Community of Practice," *Brain Traffic* (blog), https://www.braintraffic.com/insights/how-to-build-your-community-of-practice

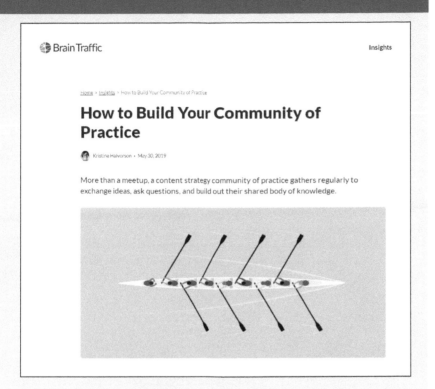

FIGURE 3.2
"How to Build Your Community of Practice," is an excellent blog post by Kristina Halvorson, founder of the Brain Traffic content consultancy (and author of *Content Strategy for the Web*) that outlines the many benefits of supporting regular meetings of content practitioners from across your company, including sharing ideas and accomplishments, and problem solving.

A content *Center of Excellence* (*CoE* for short) is an organized way to help raise the quality bar for content. How does your team define "good" or "excellent" content, beyond your heuristics? Sometimes a Center of Excellence is created by the members of a Community of Practice, although it doesn't need to be.

When you create a Center of Excellence, you might include some of these steps:

- Develop and document your content heuristics.
- Create your style guide and voice-and-tone guidelines and validate them by using content research.
- Document examples of content "dos" and "don'ts."
- Audit content to understand what content formats your team is using (videos, webinars, emails, websites and app contents, and conversational content for bots). For each of the different types of content you create, you can also document examples of excellent content, acceptable content, and content that doesn't meet your standards.
- Evaluate the current state of your content quality.
- Determine how you want to measure content impact, which can be as simple as scorecards, or as in-depth as partnering with your analytics and engineering teams to create data dashboards.
- Share examples of your content impact.

Distilling Your Content Principles

Keep in mind that your heuristics don't need to be set in stone, but ideally, they won't need to be updated or iterated frequently. The more stable you make them, the easier it will be for your content team to work with them, and for your teammates and colleagues to identify and understand them. Where a style guide is often a living, breathing reference that's frequently expanded and updated as your company's business and products change, your heuristics ought to be more of a solid list that doesn't need to change very frequently.

Identify Patterns to Inform the Principles

One great way to identify your heuristics or principles is to ask your content colleagues to keep a running list—for a day or a week or so—of their self-talk as they go about the business of creating and revising content. As your coworkers go about their day-to-day work, what's at the top of their mind, in terms of working hard to ensure the content is top-notch? Putting the customer first? Eradicating jargon? Supporting your brand voice? Simplifying the complex? Using inclusive, sensitive language? Does anything in particular from your company's brand guidelines bubble up? The building blocks of your team's content heuristics or principles will become clear when you evaluate your team's everyday practices.

Gather your team after that day or week and document what the most common elements are. Next, work to cull the list to 10 or 12 major content quality categories. Then define each one briefly, in a way that's understandable to anyone in your company and not just to the content pros.

Prioritize the Principles

Note that listing your content principles in order of priority is key. Clearly prioritizing them will help you and your content team create content more quickly. It will also help you respond to unsolicited, subjective, or just plain inappropriate feedback about your content.

Here's an example of how the prioritization of these principles may come into play: let's say you have a product manager who says, "The content is too long." (Clearly, anyone making such statements is probably new in their career and hasn't yet been made aware of the power of effective content.) Refer to your content principles when you (diplomatically) respond to such feedback. Remind your product manager that clarity and brevity are both important, but according to the team principles, clarity is *more* important than brevity. A statement like "The content is too long" is subjective. The objective, prioritized list of content principles help you respond effectively to such subjective feedback.

Or suppose that you have a product designer who says, "I don't want the words to clutter up the visual design." You can point out that clarity is an important content principle, and words are needed to make sure the user experience is understandable. That means they're essential, and far from clutter.

Examples of Content Quality Principles

Table 3.1 shows a sample list of content principles for a product team at a global tech company. Note how many of them require further explication, and how they link to wikis and style guides that go into greater depth and detail. But as a short list of easy-to-remember qualities, the principles provide a solid foundation for evaluating content—whether you're working on product content design, content marketing, website content, sales enablement content, social media, or your intranet—you name it.

TABLE 3.1   AN EXAMPLE OF HEURISTICS FOR HIGH-QUALITY CONTENT

Prioritized Content Principles	Details
Inclusive and Accessible	Our audience is diverse and global. We ensure that our content respects all cultures and gender identities and follows the inclusivity and bias-free writing guidance in the company's Writing Style Guide. For example, instead of "manpower," we use "personnel" or "workforce." We avoid slang that can be considered cultural appropriation. We also ensure that content is accessible. For example, we include alternative text on all photos and images, so customers who have low vision or blindness and rely on Braille readers can understand what is displayed on-screen. We follow Web Content Accessibility Guidelines (WCAG).
Clear	Language uses words that are familiar to the audience and avoids jargon and acronyms. Abbreviations or acronyms, if absolutely necessary, are spelled out at first mention.
Understandable and Actionable	Content helps the user know what they need to do. The next steps follow logically and are clear and complete.
Confidence Building	We try to help the customer feel confident. To create this sense of confidence, we provide words of encouragement during multistep tasks and include definitions of any potentially confusing or unfamiliar words or phrases that appear in the user interface.
Complete	Our content provides all the information and context the customer needs, at the exact time that they need it. Content is usability tested whenever possible, to help identify where any essential information may be missing.
Consistent	The content is coherent, meaning it uses consistent terminology and voice across customer experiences (including the website, app, email, SMS, advertising, social media, and other content formats like videos, brochures, e-books, webinars, and podcasts).
Empathetic	The content's tone is conversational whenever appropriate and aligns with the company brand and voice-and-tone recommendations.
Concise	Content is as succinct as possible. However, we recognize that clarity and completeness are much more crucial than brevity.
Customer-centric	Content focuses on the customer's goals and the value of the product or feature to the customer, not the business's goals. Content is scrubbed of jargon or "business-speak."
Easy to Translate and Localize	Content is simple and straightforward to translate for the languages and the cultures of our global audience. We avoid colloquialisms and slang.

Isn't it interesting to see that "findable" is a content quality element of paramount importance to so many teams? It should go without saying, but there's really no point in creating content if your audience can't find it. Following search engine optimization (SEO) recommendations, well-formed URL structure, and keyword-rich content is just plain smart. No matter what kind of content you create, make sure that your audience can find it! For more guidance on SEO, read the guidelines on Moz or Search Engine Journal's 35-step "SEO for Beginners: An Introduction to SEO Basics"[3] (see Figure 3.3). More advanced SEO best practices can be found in Google "Search Central's Search Engine Optimization (SEO) Starter Guide."[4]

FIGURE 3.3

Search Engine Journal (SEJ) is a high-quality resource on SEO best practices. While some resources about SEO make recommendations based on hearsay or trends, SEJ has been around since 2003 and is known for trustworthy, evidence-based content.

---

3   Search Engine Journal. "SEO for Beginners: An Introduction to SEO Basics," www.searchenginejournal.com/seo-guide

4   Google Search Central. "Search Engine Optimization (SEO) Starter Guide," https://developers.google.com/search/docs/fundamentals/seo-starter-guide

Figure 3.4 shows another example of content design heuristics from Ksenia Cheinman, assistant director of user experience at the Canada School of Public Service in Vancouver, British Columbia.[5] The heuristics she and her team developed are organized like Maslow's Hierarchy of Needs, with Findable and Clear forming the foundation; Connected in the center; and Human and Helpful at the top of the pyramid.

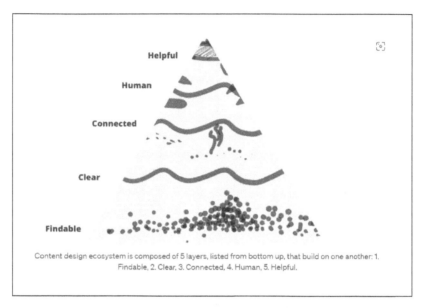

Content design ecosystem is composed of 5 layers, listed from bottom up, that build on one another: 1. Findable, 2. Clear, 3. Connected, 4. Human, 5. Helpful.

FIGURE 3.4

Another example of content principles, from the Canada School of Public Service. Findable is listed as the first foundational principle, followed by Clear, Connected, Human, and Helpful. Ksenia Cheinman is credited with developing this list of principles.

Ahava Liebtag is author of *The Digital Crown*[6] and CEO of Aha Media, a Washington, DC–area content strategy agency that specializes in health communications. She developed a short list of five qualities for

5   Ksenia Cheinman, "5 Principles of Good Content Design: How to Design Good Public Sector Content," UX Collective, https://uxdesign.cc/principles-of-good-content-design-4c55622c9919

6   Ahava Liebtag, *The Digital Crown: Winning at Content on the Web* (Burlington, Massachusetts: Morgan Kaufmann Publishers, 2013).

creating valuable digital content: Content must be findable, readable, understandable, actionable, and shareable, as seen in Figure 3.5.[7]

FIGURE 3.5
Aha Media's five-part checklist for creating valuable content, developed by Ahava Liebtag. Having just five content quality categories makes them easy to remember and helps you and your team align more easily with their recommendations. Liebtag is a recognized longtime expert in content strategy and the author of *The Digital Crown: Winning at Content on the Web.*

Under the "findable" category, Liebtag recommends including an H1 tag and at least two H2 tags; links to related content, and alternative tags for images. Under "readable," she recommends using bullets and numbered lists, and to align content with the style guide. For "understandable," she advises using an appropriate content type, such as text or video; to provide context for the audience; and to respect their reading level. For "actionable," she recommends including a call to action and to summarize what to do. For "shareable," she recommends trying to evoke an emotional response; to ask the audience to share the content; and an easy way to share it, including hashtags for social media.

---

7 Ahava Liebtag, "Must-Have Checklist to Creating Valuable Content," Content Marketing Institute, September 6, 2016, https://contentmarketinginstitute.com/articles/checklist-valuable-content/

Sharing the Quality Principles Far and Wide

Having your content principles documented—and then shared inside and outside your content team—is crucial. As you work with your product or marketing team to create content research studies, you want to make sure that every research question—and every answer that participants can choose from—aligns to your content standards or principles. This is especially important for preference tests, which are typically in multiple-choice format. You're providing a handful of answer choices for your research participants to pick from. Don't provide off-brand or off-tone response options in your content research questions, or you won't want to take your study results and implement them (or worse—you may be blocked from publishing by your brand or quality-assurance team).

For example, imagine that you need to identify the name of a new feature. You create a preference test and plan to ask a group of research participants to choose which one out of five potential feature names they prefer. If your content principles include "Content needs to be inclusive," make sure that each choice in the list of multiple-choice answers is inclusive! (For more about using language that is respectful, sensitive, and inclusive, see the "Conscious Style Guide."[8])

Aligning your research study questions and answers to your content principles may seem like an obvious step to take, but it certainly is not. Content creation is complicated, and sometimes messy. This warning is worth remembering for those times when many stakeholders are involved in content research, especially when they are from a variety of teams. Each of your stakeholders will offer their own (biased) suggestions for potential answers to be included in a study. Unless each of these people is a content creator well-steeped in your style guide and writing voice and tone, you need to be on the alert to keep your research study questions "clean" and free of off-brand or otherwise inappropriate answer options.

More practically, you also need to be mindful when creating your research studies of any constraints or limitations related to content publishing. For example, if you create content for social media, you need to be careful to keep your social media posts short and sweet. If you're creating a webinar to be promoted over social media, you

---

8   Karen Yin, "Conscious Style Guide," https://consciousstyleguide.com

don't want any webinars with titles that are 25 words long. (Some colleagues refer to very, very long content asset names as Fiona Apple–inspired titles, as she holds the world record for longest album name.) Therefore, if you're running a content test to identify an appealing webinar title name, make sure that any name contenders included in your study are sufficiently short and workable.

This length constraint may be obvious to you, as a content pro, or to your production team, or the social media community manager. However, this kind of detail may be unknown to your Chief Marketing Officer or other colleagues who provide solicited or unsolicited contenders to be included in a naming research study.

As mentioned previously, ideally your list of content principles will be short and accompanied by definitions that make the principles feel concrete and relatable. (You can also include "Do this" and "Don't do this" content examples for each.) You won't be able to explain each concept in detail; it's perfectly okay to refer to additional resources for more information.

Content Voice-and-Tone Guidelines

So you have your handy short list of the Most Important Content principles defined, prioritized, and documented. Terrific! But what about voice-and-tone guidelines? They're another essential element to have in place before you get rolling with content research. Discussions about which words are "on brand" or "off brand" or "aligned to the company voice and tone" or "totally not aligned to our voice and tone" will come up frequently as part of your content research work. And, just as you don't want to conduct content research by using any words or phrases that don't align to your content principles, it's best to avoid doing any content research with words or phrases that don't align with your voice-and-tone guidelines.

The Differences Between Voice and Tone

The first step is to define voice and tone, and the nuances between the two. A helpful analogy for explaining the difference between the two has to do with weather (at least, before climate change). Here's how it goes: *voice* can be thought of as analogous with climate. It's relatively predictable, and that predictability is something that you can usually count on. (For example, when you live in Phoenix, you typically need a sun hat, but you don't need to own snow boots.) Aligning content to your brand voice makes content coherent and consistent. As a practical example, when you read content from Starbucks, the voice of its content can always be described as "passionate about high-quality coffee." The content is consistently conveying excitement about great coffee, every single time you visit a store or read their website or app.

Tone, on the other hand, can be thought of like the daily weather forecast *within your climate*. It can vary, but because it's dictated by the climate, it doesn't veer too far outside the expected norms. When describing its cold-brew coffees on its website, Starbucks chooses words like "nuanced," "smooth," and "extraordinarily uplifting."[9] Each of these descriptions falls under the tone umbrella of "elegance," which ladders up to and supports the "passionate" company voice.

The tone they use to describe other product lines—their Frappuccinos or whole bean coffees—differs from how they describe their cold-brews. Yet they still support the content voice that's centered on passion. Here's another example shown in Figure 3.6, for the Single-Origin Guatemala Casi Cielo whole bean coffee. Enthusiastic descriptions like "most beloved," "finest possible expression," and "not to be missed" also convey a tone of elegance and quality, which again ladders up to the "passionate" voice.[10]

---

9 Starbucks,"Cold Coffees: Cold Brews," www.starbucks.com/menu/product/2122616/iced?parent=%2Fdrinks%2Fcold-coffees%2Fcold-brews

10 Starbucks, "Menu: Whole Bean," www.starbucks.com/menu/product/4/whole-bean

FIGURE 3.6

The Starbucks website includes evocative descriptions of its coffee drinks and coffee beans. In this example of Starbucks' premium single-origin whole bean coffee, descriptions like "finest possible expression" and "select farms" convey a tone of elegance.

Something to know about voice and tone: determining which words and phrases are on or off brand can quickly devolve into splitting hairs. What tends to be helpful is providing many examples of words whose tone is on-brand and appropriate for your content creators, and many, many examples of words whose tone is decidedly *not* aligned to your voice and tone. If you have a brand manager or other arbiter of all things related to product messaging, ask for their input and assistance.

> **PRO TIP** DEFINE CONTENT VOICE BY WHAT IT IS *NOT*
>
> Another helpful way to think about voice and tone is to consider which words do not work to describe the tone of your content or are words that go too far. Here are some examples: Your voice may be helpful, but not solicitous. It may be playful, but not silly. It may be confident, but not brash. When it comes to helping you and your team create great content, defining the words that do not accurately describe your content voice can be just as helpful as the words that do define it.

How Content Research Can Inform Voice Guidelines

Just as you want to make sure that everyone involved with content research—like product managers, user researchers, product designers, and software developers—is aware of your content principles, it's helpful to make sure that they are also aware of your content voice-and-tone guidelines. You can include a link to your team's voice-and-tone guidelines from your content research template and each research study plan. That way, content research can easily align to content research tenet #1: Don't conduct research with any content options you shouldn't publish because they don't align to your content, brand, or voice-and-tone guidelines.

Ponder this for a minute: How were your company's voice-and-tone guidelines created? Chances are they were developed by your branding team. While your brand team is smart and experienced, you can probably bet that the voice-and-tone guidelines weren't vetted and validated by using content research. To the extent you're able to do so, your voice-and-tone guidelines can (and should) be evaluated using content research and be updated to reflect the content insights you uncover with your content studies. If research shows that specific words that are in your voice-and-tone guidelines are roundly disliked by or unclear to your research participants, let your brand team know and work to collaboratively update your voice-and-tone guidelines accordingly.

PRO TIP WITH RESEARCH INSIGHTS, BE AUDIENCE-SPECIFIC!

A bit of advice: Be careful to note which specific segment of your audience you're referring to when you refer to content research insights. Words that work for people who work at large enterprise-size companies with tens of thousands of employees may be quite different than those that work for entrepreneurs.

Long story short, you may have an opportunity to improve upon and update your company's voice and tone—and, by extension, its writing style guide—through content research insights.

Cite Content Research in Your Style Guide!

If you conduct research that provides additional insights to the current guidance in your voice-and-tone guidelines or style guide, and you then go ahead and make edits or updates to your style guide, be sure to include a link to your content research study plans and study results, along with the date. You can use an icon—perhaps a checkmark—or simply mark the margin of the applicable style-guide entry with "validated" or "supported by research." This will bring more awareness of content research to everyone at your company who uses the writing style guide—including NCFs, or non-content folks—like product managers, visual designers, and software developers. It can also generate more excitement for content research among your content team, when the NCFs inquire about content research.

To further support the practice of content research at your company, add your email or other contact info to your style guide and content design library and encourage others to reach out to you to learn more about content research.

Making It All Work for You

Each content team is at its own stage of maturity. You may work as a content team of one for a startup and have just a few entries in your content guidelines and voice-and-tone recommendations so that you need to start from scratch to build your list of content principles. Well-established and staffed content teams will likely have a comprehensive writing style guide and voice-and-tone tenets that have been built upon for years, along with documented and distributed content principles. Most companies will fall somewhere in the middle.

Don't beat yourself up if your resources are limited! Instead, look upon content principles and guidelines as smart tools to have in your content pro toolbox. If your content team's principles aren't yet distilled and documented, make doing so a priority goal for your next business quarter or the next six months. Chances are you'll find them to be sanity savers and time-savers that make all your work go more smoothly and quickly, and that includes your team's content research efforts.

CHAPTER 4

Evaluate the State of Your Content

Once you've established your team's content principles and which ones are most important, you've set the stage to evaluate the state of your content. How do you feel about the quality and effectiveness of the content you and your content team are creating? And how and what do you and your team confidently know about it? *How do you know what you know?*

Do you currently have a standardized way of evaluating content, through scorecards, performance dashboards and metrics, or other methods? Do you use a method like NPS (Net Promoter Score[1]) that's notorious for being a challenging way to measure the effectiveness of content and customer experiences, but somehow finagled its way to become a popular metric for many product teams? Do you evaluate your competition's content to compare and contrast it with your organization's?

If you do evaluate content, is it done regularly or only periodically or sporadically? Is your company's MIC (Most Important Content)—like your most important landing page or an often-used section of your app—high quality, while much of the rest of the content is just so-so? Or maybe you truly have no idea whether it's high quality, because a lot of your content hasn't been touched in ages (a very common problem for understaffed content teams).

PRO TIP CONTENT SLEUTHING

There's quite possibly more content evaluation happening than you're aware of—even though you're on the content team. Evaluating content requires some of what I refer to as *CSI: Content*, or *Crime Scene Investigation: Content.* You're going to need to ask around and talk to people in roles both inside and outside your immediate content team to get a full picture of how your users are engaging with and reacting to content.

Methods for Evaluating Content

Your approach and methods for evaluating the overall state of your content will vary, based on several factors. If you and your team are currently looking at the state of your content on a regular basis and

---

1   Jared M. Spool, "Net Promoter Score Considered Harmful (and What UX Professionals Can Do About It)." Medium, https://jmspool.medium.com/net-promoter-score-considered-harmful-and-what-ux-professionals-can-do-about-it-fe7a132f4430

using a framework that's working well for you and your stakeholders, count yourselves lucky! If your team has never had the time or tools to take stock of your content quality, then check out the many options and choose what will work best for you, given your resources (people and time) and budget (see Table 4.1).

TABLE 4.1 BEGINNER-LEVEL CONTENT EVALUATION TOOLS

Content-Measurement Method	Tools or Platforms Used	Definition and Pros and Cons
Rating Scales	Simply your own judgment	Rating-scale measurements are a quick "which way is the wind blowing?" evaluation of content quality, usually on a simple scale of 1 (needs a lot of work) to 10 (excellent). Pros: Quick and simple Cons: Subjective and not rigorous
Scorecards	You can create your own scorecard framework that's based on your content team's content-quality principles or heuristics. For example, if you have five core content principles, you can rate a specific content experience against each one. If you decide each principle can score a maximum of 10 points, then the maximum score would be 50.	A subjective evaluation of a specific content experience or body of content, judged against your content team's vetted, defined content standards or principles. Scorecards provide an easy-to-understand score out of a total possible number (like 50 or 100). The practice of using scorecards is referred to as content "grading," as you can convert the total scores to easy-to-understand "grade" equivalents like C+ or A-. To use a scorecard to evaluate a body of content (like a collection of three landing pages, or a series of five emails), take the average score of the individual pieces of content or content assets. Pros: Relatively quick and easy to use Cons: Subjective. Like rating-scale measurements, you can somewhat reduce the subjectivity of the measurements by asking several teammates to evaluate the content experience in question using your scorecard framework, and then calculate the average score.

continues

TABLE 4.1   continued

Content-Measurement Method	Tools or Platforms Used	Definition and Pros and Cons
Readability and Clarity Measurements	Readability and clarity can be measured with the help of several apps or tools. This is not an exhaustive list. Hemingway app Readable app Grammarly app Flesch-Kincaid formula Editor (a tool built into Microsoft Word)	These tools use an algorithm to measure content's overall readability. Pros: The Hemingway app, Flesch-Kincaid Reading Ease formula, and other similar tools are helpful for flagging when content is overly complex and wordy and in need of clarification and revision from content experts. Cons: They're not perfect. Flesch-Kincaid, for example, uses a complex algorithm that involves the number of words per sentence and the number of syllables in each word, which may or may not be an accurate way to measure how easy it is to understand the content. Flesch-Kincaid is *not* a great tool for user experience (UX) content, because it is intended to measure full sentences, and its accuracy increases when full sentences are being evaluated. Another drawback to readability measurements is that teams may become overly fixated on using these tools to achieve a very specific level of clarity, such as the sixth- or eighth-grade reading level that many plain-language experts recommend for digital content. (See Chapter 2, "Leverage the Power of Clear Content.") Teams outside of content, like product management, can become very attached to readability metrics. This fixation on reading level can sometimes detract from the other work that content teams need to accomplish. (See Chapter 3.)

Rating Scales

For rating-scale measurements, you identify a scoped, specific content experience—for example, a single email, a landing page, or a specific multistep user or customer flow. You then subjectively rate it on a scale of 1 to 10, with 1 being the lowest quality and 10 being the highest.

Subjective is the key word here, and it's exactly why rating scales are novice-level metrics. The subjectivity makes them less useful and reliable than other, more advanced types of measurement. To make rating-scale measurements more useful, you can ask several coworkers to rate the content and then take the average rating.

If you've taken the time to develop content heuristics or quality principles, you can take the rigor of your content evaluation a step higher by using those heuristics in your measurements.

Scorecards

You don't need a hefty budget or fancy tools to understand the overall state of your content. While basic tools don't provide the depth or breadth that the more advanced ones do, sometimes all it takes to judge your content is a simple scorecard framework, and a bit of time to devote to evaluating your content.

Why Use Scorecards?

One way to create a scorecard is to take your content principles or heuristics and assign a point value to each. Say you have (a convenient number of) 10 content principles, and the specific piece of content or content experience you want to evaluate is a new customer onboarding user flow. You can evaluate the onboarding experience using those 10 content principles. Score each principle on a scale of 1–5, for a potential maximum total of 50 (or 1 to 10, for a potential total of 100).

This type of content evaluation is helpful as you embark on content research so you have a feel for how your content is faring, and you can use this as a benchmark score. Re-evaluate and rescore the content you iterate thanks to content research, and you should have a "before-and-after" improvement story to tell. (These stories tend to go over especially well with senior leadership, which can also bring more energy to your content testing program.)

Sample Content Scorecard and "Grade"

Figure 4.1 shows a simple example of a content scorecard—one used to evaluate user experience writing, or *content design*. Obviously, there is quite a bit of subjectivity involved. If you asked your content director to rate a specific piece of content, the score they determine may be quite different from how a junior content creator might score it. And that junior content creator's score might be quite different from that of the person who did the hard work of creating the piece of content in question. However, to work around the subjectivity involved in score carding, you can obtain an average. Ask several colleagues to individually evaluate a content asset and then take the average of those scores.

TRACKING PROGRESS WITH SCORECARDS

You can help generate buy-in and support for your content research program by documenting how research improves content quality scores. To do so, use your scorecard framework to measure the content experience you'll be conducting research on, to first get a baseline. Then conduct your research. Next, update your content to reflect the insights you discovered through conducting the research. Score your content again. How do the two measurements differ? Chances are, you're improving your content substantially, by uncovering insights about your customers' preferred language and words and updating content accordingly. Communicate this positive difference to your whole team!

(This is easiest to do with existing content. If you're creating brand-new content, you can score or rate the early first drafts of content you create before conducting research. Compare the scores of the initial content to the scores of the final version.)

You can also create support and buy-in for your research program by setting content score improvement goals for your team, using a spot audit framework. For example, say you're managing content for your company intranet, and there are 20 different content sections. It would be very time consuming to evaluate and score *all* the content in all 20 sections. Instead, you can choose a smattering or sampling of content. You could randomly select 5 of the sections and score that content—or a smaller portion of all 20 sections. Calculate the average of those scores for your baseline.

Then set a goal for your team to increase the score by a certain percentage, by a specific date. Use content research to improve the content, and then re-evaluate and score the content again. You should see an improvement over time and be able to quantify the impact that your content research and content team is making!

Scorecard for Content Design Team

DATE: January 12, 2023

FIGMA/DESIGN FILE LINK: figma.com/xyz123

NAME OF EXPERIENCE: Product XYZ purchase experience

Product Team Contact Info

CONTENT DESIGNER: Ruth Gutiérrez

PRODUCT MANAGER: Mike Evans

PRODUCT DESIGNER: Carrie Noonan

SCORECARD REVIEWER: Trudy Fowler

Scoring Guide

EXCELLENT: 100

Approved

GOOD: 80-99

Approved with minor updates

FAIR: 60-79

Not approved; rework needed

POOR: 0-59

Not approved; extensive rework needed

SCORING CRITERIA	SCORE (0 TO 10)	COMMENTS/ISSUES FOUND	RECOMMENDATIONS
Clarity	6	Some jargon and acronyms are included.	
Customer centricity	6	Explanatory content is long and technical. Some words and terms need definitions added. Payment steps include complicated tax information.	Meet with legal team to understand what tax and payment details could be omitted or revealed using progressive disclosure.
Accessibility	10		
Actionability	10		
Consistency	8		
Empathy	9		
Inclusivity/sensitivity	8		
Conciseness	8		
Branding/voice and tone	3	Tone could be warmer and more welcoming.	Expand the introductory section to thank the customer for choosing Brand XYZ.
Completeness	8		
TOTAL	76		
Recommendation: Not approved; rework needed			

FIGURE 4.1

A sample content scorecard, leveraging content heuristics, that rates content against each of those heuristics as poor, fair, good, or excellent.

Readability and Clarity Measurements

Measuring readability and general clarity—as gauged by digital apps and tools like the Hemingway app, the Flesch-Kincaid algorithm, or the Editor feature built into Microsoft Word—are other basic but helpful ways to evaluate your content. See Chapter 2, "Leverage the Power of Clear Content."

While readability tools are easy to use, they can also become stumbling blocks. If your content team uses them in a rigid way, they can and will slow you down. For example, if your content quality principles state that content clarity should score no higher than a sixth-grade reading level, then content creators will get hung up while writing and constantly retake clarity measurements, sometimes multiple times for one paragraph, to reach that specific clarity level. (Quality principles are better off being recommendations instead of rigid rules.)

Instead of using clarity measurements in a strict way, give your content team the power of using their best judgment. You don't want your team to be spinning its wheels to ensure that every piece of content that you create achieves a predetermined grade level or score. You also don't want product managers or other stakeholders to similarly become obsessed with scoring in this way.

That said, taking the time to measure the clarity of your writing for the first time can often be shocking for a content team. You may think your writing is clear and crisp, only to discover that it is hitting a graduate school reading level. Use these tools sparingly and wisely! Remember, if you have content heuristics, you have a whole list of other content quality measures to take care of.

Intermediate-Level Content Measurements

Evaluating content with industry-standard measurement tools and other methods like heat maps, which are more sophisticated than homegrown scorecards, can provide credibility for your content team and a variety of valuable perspectives on content effectiveness. They also can reduce the subjectivity inherent in beginner-level measurement tools like scorecards, and therefore help build greater respect for your content team. As seen in Table 4.2, sentiment analysis tools and heat maps reflect customer behavior, which makes them invaluable tools for content pros. SEO (search engine optimization) tools and website quality platforms like Siteimprove can also help you dramatically improve your customer content experience.

TABLE 4.2 INTERMEDIATE-LEVEL CONTENT MEASUREMENT TOOLS

Content-Measurement Method	Tools or Platforms Used	Definition and Pros and Cons
Sentiment Analysis Tools	Net Promoter Score (NPS) System Usability Scale (SUS) System Ease Score (SES) Others: IFS Customer Experience Management (formerly Customerville), Qualtrics' CustomerXM	Online survey tools with a templatized set of questions Net Promoter Score was developed primarily as a brand or customer sentiment tool and is not intended to measure content effectiveness. It measures how likely people *claim* they are to recommend a product or service to someone else.[2] It is much more simplistic than SUS or SES. (Remember, what people *say* they will do often differs from what they actually do!) SUS[3] and SES provide deeper insights than NPS since they ask more—and more detailed— questions. However, their intent is to measure peoples' *perceptions* of usability—not actual behavior itself. Pros: SUS and SES provide more in-depth snapshots of the state of your content and design. They can both provide baselines of content at a specific point in time and can then be rerun at a later date (say, a business quarter or six months) to gauge change in quality over time. Cons: SUS and SES can be time consuming to implement. NPS is a controversial—some might even say disparaged— measurement tool, especially for user experience content. Proceed with caution!

continues

---

2   Jared M. Spool, "Net Promoter Score Considered Harmful (and What UX Professionals Can Do About It)," Medium, https://jmspool.medium.com/ net-promoter-score-considered-harmful-and-what-ux-professionals-can-do-about-it-fe7a132f4430

3   Jeff Sauro, "Measuring Usability with the System Usability Scale (SUS)," MeasuringU, https://measuringu.com/sus/

TABLE 4.2 continued

Content-Measurement Method	Tools or Platforms Used	Definition and Pros and Cons
Heat Maps	Crazy Egg Design Flight (with Figma) Microsoft's Clarity app	Pros: They tell a clear story of where customers' attention is going. They're actionable and easy to use. Cons: Heat maps show where a customer's attention is, but don't tell you what the customer is thinking. In this way, they tell only a partial story. When you use heat maps, additional research is often needed to fully understand what's going on in the minds of your users.
SEO Keyword Research Tools and SEO Analysis Tools	Google Trends Google Search Console Semrush Moz Pro, Moz Premium Screaming Frog BrightEdge STAT	Pros: Google Trends is free and easy to use. Moz and Screaming Frog provide free trials you can use to easily and quickly evaluate your content. Cons: Semrush and BrightEdge are expensive and have steep learning curves. Also, SEO is a dynamic field that continually changes, so it can be tough to stay on top of best practices.
Accessibility and Other Website Quality Tools	Siteimprove	Pros: Siteimprove will score your site on how well it aligns to WCAG accessibility standards (A, AA, or AAA), and it provides details in an easy-to-use dashboard view. It also offers evaluations of how inclusive and sensitive your content is. You can scan your site's content accessibility with their tool: www.siteimprove.com/platform/accessibility/. Cons: Siteimprove is so thorough and comprehensive that it might keep you up at night as you think about all the improvements you need to make to your content.

Sentiment Measurements Like NPS, SUS, and SES

Metrics like Net Promoter Score, System Usability Scale, and System Ease Score use a series of survey questions to evaluate how your audience members state they feel about the content experience. Each response is typically measured along a scale of 1 to 5, or 1 to 10, with 1 being the lowest score possible and 5 or 10 being the highest. Each of these metrics aims to measure the overall customer experience, with SES and SUS looking at both content and product design. To use these tools to understand the state of your content, you'll need to tease out the questions that pertain to content—and only content.

NPS is the simplest of these three measurements. It aims to evaluate the degree to which people *say* they're likely to recommend a service or product to friends or family. You've likely encountered an NPS survey without knowing what it's called. Whenever you're asked to rate your experience with a customer service representative and whether you'd recommend the company or product to a friend or family member, that's a Net Promoter Score survey in action.

While NPS is a commonly used measurement for sentiment, it's far from being the most reliable or actionable metric for content creators to use to measure content effectiveness. (Many UX practitioners would recommend avoiding it, as it can be a mercurial measure, with scores rising and dipping with no direct correlation to content quality. You certainly don't want to use NPS as a metric for your annual performance review.)

All three measurements—NPS, SUS, and SES—are survey-based metrics that can be time consuming to set up and evaluate. On the plus side, you can use SES or SUS to take a baseline measurement, say, at the beginning of the calendar year or start of a business quarter. Then you can repeat the survey every month, or over whichever timeframe that you decide. The expectation is that the more you improve your content and customer experience, the greater the score will improve. (Because they're complex and time consuming, if you wish to evaluate your customer experience using SUS and SES, you may want to involve your user experience research team.)

The following 10 questions are asked when measuring System Usability Score. You'll notice that the questions refer several times to the word "system." You can replace the word "system" with the specific product, feature, or customer experience that you're measuring, so long as you avoid jargon and use a customer-friendly description of that experience.

For example, internally at your company, you may refer to the experience that new customers have with your product or service as "onboarding." However, that's not likely the same language that your customers use. Customers are probably more likely to think of that experience as "getting started"—so that's the language that should be used in your SUS survey questions. SUS was developed by John Brooke and popularized by Jeff Sauro, founder of MeasuringU in Boulder, Colorado, and co-author (with James R. Lewis) of *Quantifying the User Experience.*[4]

1. I think that I would like to use this system frequently.
2. I found the system is unnecessarily complex.
3. I thought the system was easy to use.
4. I think that I would need the support of a technical person to be able to use this system.
5. I found the various functions in this system were well integrated.
6. I thought there was too much inconsistency in this system.
7. I would imagine that most people would learn to use this system very quickly.
8. I found the system very cumbersome to use.
9. I felt very confident using the system.
10. I needed to learn a lot of things before I could get going with this system.

The value of using a more time- and effort-intensive evaluation tool like SUS or SES is that their results are more objective than those from basic tools like scorecards. You can also use SUS and SES with a smaller number of participants and still get valuable, actionable results.

Heat Maps

Heat maps can show where specifically within a web page or app screen your users are hovering their finger or mouse, or where exactly they are clicking or selecting content to interact with. Heat maps can also be an effective tool for better understanding how clear and

---

4   James R. Lewis and Jeff Sauro, *Quantifying the User Experience: Practical Statistics for User Research* (Burlington, Massachusetts: Morgan Kaufman Publishers, 2012).

effective your content is (see Figure 4.2). Are customers easily moving from step to step in your customer flow, or are they apparently getting stuck on one or more of those steps? Is irrelevant information distracting them and preventing them from doing what they need to do—or what you want them to do—when using your content? Heat maps can help you understand your customer more deeply.

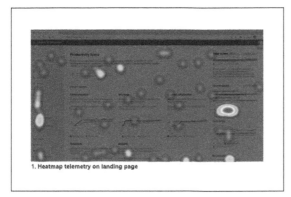

1. Heatmap telemetry on landing page

FIGURE 4.2
This heat map shows that the participant's attention is focused on a landing page at the mid- to lower part of the left navigation (the green teardrop shape with light-blue outline) and at the middle-right (the red oval surrounded by yellow, green, and light blue).

SEO Tools

Search engine optimization (SEO) is the art and science of making sure that your content is easy to find when people do a web search for your business, product, or other terms related to them. It's a complex, continually evolving field. Dozens of tools exist to help you understand what words and terms ("keywords") people are searching for on Google, DuckDuckGo, Bing, and other web browsers. Many more tools exist to help you improve the SEO "health" of your app and website.

Like web analytics, SEO is a complex topic worthy of tome-size books and pricey, weeks-long certification programs, and is too complicated to cover thoroughly here. If you want to explore more about SEO and understand more about keywords and their importance to your content ecosystem, Moz offers an excellent and free "Beginner's Guide to SEO."[5] (See Chapter 3, "Identify Your Content Quality Principles" for more details on getting started with SEO.)

---

5 Moz, "The Beginner's Guide to SEO: Rankings and Traffic Through Search Engine Optimization," https://moz.com/beginners-guide-to-seo

One of the most basic SEO tools used by content pros is Google Trends. It reveals which specific words or phrases are most popularly used by the public. In this way, Google Trends (as seen in Figure 4.3) can help you quickly gauge whether specific words you're using are familiar to your customers, or if you're using unfamiliar or outdated terms.

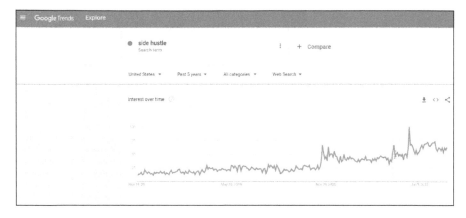

FIGURE 4.3
Google Trends is a free tool that gauges the public's familiarity and interest in specific words and phrases over time.

Screaming Frog[6] is another fascinating SEO-focused content platform, although it probably has the strangest name for an app, ever. Screaming Frog is what's called an SEO "spider" that "crawls" your website or app at the code level to evaluate its structure and content. (See Figure 4.4 for a sample content audit conducted using Screaming Frog.) Screaming Frog will flag exactly where your app or site has broken links, as well as if you have missing elements that Google and search engines need—like page descriptions—to decide whether your content on specific topics deserves to appear at the top of search results, in the middle, or down at the bottom. The good news around SEO is that if you create high-quality, truly unique content that serves a specific customer need (or "intent," as they say in the SEO field), you're ticking off the most important SEO best-practice checkbox.

---

6   Screaming Frog, https://www.screamingfrog.co.uk/seo-spid

FIGURE 4.4

This partial audit of the University of Washington website was done using the Screaming Frog content audit tool, available from **screamingfrog.co.uk**. It reveals which website pages include well-written page titles and meta descriptions, and where they're duplicative, too long, too short (according to Google standards), or missing.

Advanced Content Measurements

Don't be daunted by advanced content metrics. You can think of them simply as more durable measurements than the beginner or intermediate methods. Getting established with tools that measure advanced metrics can be expensive, both from a time and budget perspective. But if you manage to successfully lobby for them, they're well worth the investment. They can help you further foster respect for your content team, because these measurements clearly show that when you invest in high-quality content, customers are better able to do what they need to do and accomplish their Jobs to Be Done— which translates into happier customers and a stronger bottom line for your business. See Table 4.3.

TABLE 4.3   ADVANCED CONTENT MEASUREMENT TOOLS

Content-Measurement Method	Tools or Platforms Used	Definition and Pros and Cons
Behavioral Metrics	Tools and dashboards that measure customer behavior, such as task completion rates GA4 (also known as *Google Analytics*) Adobe Analytics Custom dashboards created using Power BI or other "home-grown" dashboards developed by analytics teams Contentsquare	Pros: Provide a fascinating view into customer behavior. Help you spot trends and changes in the way that your customers are interacting with your content. Can be very eye opening to senior leadership and help you advocate for additional content staffing. Cons: Can be expensive, and some have somewhat steep learning curves, depending on how sophisticated you want to get with their bells and whistles.
Combined Sentiment and Behavior Metrics	Google's HEART framework: Happiness Engagement Adoption Retention Task Success	Pros: Shows both the *what* (what customers are doing) and a bit of the *why* (how they're feeling as they're using a product or feature). Tools that show combined sentiment and behavior metrics help highlight when customers are successfully completing a specific task or Job to Be Done, but they are not pleased about the experience (due to reasons like feeling frustrated or unconfident while using the website). Tools that combine sentiment and behavior metrics also show the opposite: when customers are unable to complete a task but that inability doesn't seem to bother them. Cons: Like all metrics, it's hard to know 100% what the story is since you aren't observing the customer or outright asking them questions. If a customer can't complete a task, is it because the task is truly too complex and confusing, or is it because someone rang the doorbell, or their toddler distracted them? If they're happy while completing a task on your site or app, is it because they love that it's easy to complete that task, or did they just hear some completely unrelated but wonderful personal news?

Digital data platforms including Google Analytics and Adobe Analytics measure a variety of customer behavior that can be directly related to the content experience. (If you use a content management system, you may be able to use its dashboard to view customer behavioral data and may not need or want a sophisticated analytics platform like Google's or Adobe's.) Google Analytics, Adobe Analytics, and similar platforms can display data visualizations that reveal customer engagement and behavior, which can be an excellent way of evaluating the effectiveness of UX writing or content design. They include "click-path" data, which reveal step-by-step how your customers are navigating through your website or app.

You may have created what *you* think of as a linear, easy and clear, step-by-step customer experience, but the click-path data can reveal just how differently your customers navigate through it than you might expect. When customers click or tap around a site or app in a circuitous or back-and-forth way that signals they're having difficulty finding what they want or completing the steps they need to, it's sometimes referred to as *ping-ponging*.

For example, if you work in ecommerce, you might need to ask customers to create an account on your site. Let's say that it's a four-step process, each process with its own screen on your app or website:

Step 1=Enter name and mailing address.

Step 2=Enter email address.

Step 3=Enter credit-card information.

Step 4=Confirm.

Google Analytics and Adobe Analytics can show how many customers move through these steps in the expected order, from Step 1 to Step 4. They can also show how many people get stuck and "abandon ship" at which specific step. They'll also show how many people move a step backward, then forward, then backward again, ping-ponging and seemingly unable to make up their minds or fill out the (seemly straightforward and simple) form you're hoping they'll complete.

You can filter these dashboards by day, hour, or minute. Therefore, they'll help you determine whether any new or updated content added to your website or app prompts new or different behavior from your audience. Was a clarified explanation that was added on Wednesday morning able to boost form completion rates? These sophisticated tools can help you tell the story of how content, when continually iterated and improved, is directly responsible for customer success.

So you created some content—a specific landing page, or section of your app—that your analytics platform shows is popular and getting a lot of visitors. Don't fall into the trap of assuming that it's therefore your best, highest-quality content. Sometimes, a web page gets traffic, clicks, and engagement simply because it's the easiest to find. Findability does *not* equate to quality. The content may be popular due to any number of factors: because there's a neon purple call-to-action button on your home page that links to it, or it's easily found by your customers due to your website or app's navigation, or maybe Google has blessed it and ranks it high for search results.

The sad and somewhat scary truth is that often the digital content that's the easiest for customers to find is the oldest and most outdated (and sometimes even inaccurate and sloppy due to neglect). Like the most popular girl in high school who can't get a date to the prom because everyone assumes she's already been asked, it's easy to assume that your content that's getting viewed the most must be high quality and doesn't need any attention.

Here's the story: Google and other search engines deem the age of your website to be of utmost importance when deciding if they want your content to rank high in search results and be easy to find by people who are doing web searches. In fact, the age of your website domain is the number-one "ranking factor" for SEO. Older websites and older web pages, because they're affiliated with stable companies that have been around for years, are looked upon more favorably by Google and other search engines. On the other hand, Google is skeptical about brand-new companies and ranks their sites lower, at least until

Behavioral Metrics Like Task Completion

Here's an example of a basic custom dashboard that provides an actionable "funnel" view of multistep user experiences, or user flows. The funnels in this dashboard measure task completion rates (TCRs), which reflect what percentage of users successfully move from one step of the flow to the next (see Figure 4.5). For example, if it takes seven steps for a customer to set up their software, the dashboard shows the percentage of users who are succeeding at each of those seven steps for a specific date or date range. It also very clearly shows what the "fall-off" is at each step (see Figures 4.5 and 4.6). These behavioral metrics are in a way more valuable than sentiment metrics because they feel clearer and very actionable. If 50% of customers are getting stuck on step 3 of that 7-step flow, then the content and product design teams know they need to do some due diligence and revisit the content and the design at that step to make sure it's clarified and improved.

lots of people engage with their content. (And getting traffic to a new site is challenging when potential customers can't find your website in the first place!)

Companies launch new products and features all the time, and those products and features need fresh, brand-new content pages to be published to go along with them. The net result of continually creating that spanking new content for fresh products is that the oldest content on your site may not get any care and attention for years—if ever. Sound familiar? This is often referred to as *content rot*.

The oldest content pages get respect and higher rankings from Google and get seen by your customers. But chances are, since they get a lot of traffic, they're less likely to capture the attention of you and your content team. Popular content is often rife with inaccurate information, maybe even details about products and services that your company has stopped offering. (Might your Help and FAQ pages and About Us content fall into this category?) Content needs regular care and feeding to thrive.

It's easy to make the mistake of assuming that popular content equals good content. Don't be fooled! Your oldest content pages—which are often the easiest for customers to find using web searches, thanks to Google's ranking algorithm—are likely to be neglected and outdated to an embarrassing degree. Content governance—which includes regularly reviewing and updating content—is often an overlooked content best practice, and it's essential for a strong customer content experience.

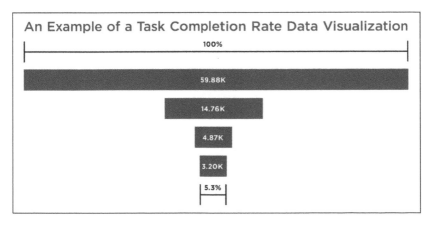

An Example of a Task Completion Rate Data Visualization

100%

59.88K

14.76K

4.87K

3.20K

5.3%

FIGURE 4.5

A sample user experience "funnel" view that shows a four-step user experience with a very low overall success rate of only 5.3%.

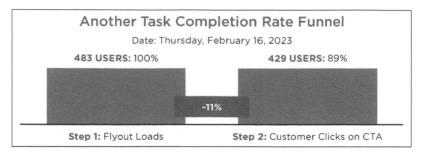

FIGURE 4.6

A horizontally oriented content funnel view showing a two-step user-experience flow with an 11% drop-off from Step 1 to Step 2.

Other behavioral metrics include time on site (how many minutes or seconds users spend navigating through your site or app), click-through rate (what percent of site visitors click or tap on a specific link or call to action), and revenue-related metrics like purchases (for ecommerce sites). Behavioral metrics on their face can feel more valuable than sentiment metrics. It's hard to quibble about customers clicking on links and buttons or completing a purchase. However, it bears repeating that just because content is clearly popular, *that doesn't necessarily mean it's the most valuable content.* It may be because the landing page your audience interacts with most often is simply the easiest to find, due to search engine optimization. Your oldest content by default may be the easiest to find, because it's had more engagement and therefore search engines will typically make it easier to find than brand-new content. Keep in mind that you as a content pro may be aware of this scenario, but you may need to help educate your colleagues who don't work in content and help them understand it.

Combined Sentiment and Behavioral Measurements

Evaluating how your users feel (their sentiments) while they're interacting with your content can be helpful. Evaluating how the people are behaving when interacting with your content is terrifically helpful. But why not both? It's incredibly valuable to know how your users or customers are feeling *as they interact with your content.*

Truly, the Holy Grail of content measurement is metrics that combine the *what* (what people are doing—their behavior) and the *why* (why they apparently feel the way they do while they're interacting with your content). When you have both measurements, that can help minimize the risks inherent in each measurement. For example, you may have a home page where you can see, using click-through-rate data, that most customers are selecting the primary call-to-action (CTA) button and not interacting with many of the other links on the page at all. But are they happy? Are they frustrated by too much or not enough detail? What's going on in their minds that might affect whether they return to your site in the future or remain a loyal customer of your product or service? Combining both the *what* and the *why* metrics simply give you a more complete (if still imperfect) picture of the customer content experience.

One tool that can help you understand both the what and the why was created by Google. It's an amalgam measurement called the *HEART framework* (see Figure 4.7). Here's what it measures:

- Happiness
- Engagement
- Adoption
- Retention
- Task success

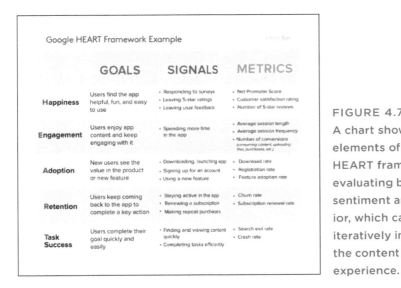

FIGURE 4.7

A chart showing the elements of Google's HEART framework for evaluating both user sentiment and behavior, which can help you iteratively improve the content customer experience.

Hunting for Yet More Content Facts

No matter what the size of your company—startup, medium-size business, or global enterprise—chances are, there's been research completed in your company that relates to content performance that you and the rest of your content team haven't seen yet. Think of all the teams you cross paths with: brand, product management, user research, and any and all marketing divisions (to make sure that you're covering all the personas and audiences and each of the different products you create content for). Ask your content-team colleagues to reach out to the people they know on these teams and ask them for any information or reports they can share that might benefit your content team and shine a spotlight on how customers are reacting to content. Prepare to be surprised!

If your company pays for continuing education such as by sending employees to conferences (for example, Adobe Summit), they can be excellent sources of "industry standard" digital experience best practices and metrics and guidance that can also help you and your content team be more effective in your roles. Product management and engineering teams often have larger budgets for such events than content or design teams. Ask around and see how your content team may benefit from the knowledge shared at these events. Similarly, if you and your content team are fortunate enough to attend content-focused events like the Button Conference, let your product colleagues know about what you learned.

Competitive Analysis

When you ask your colleagues for information related to customer research, you'll possibly be sent some competitive analysis reports, especially from the product team. These reports can be gold mines of information and can also be quite inspiring for your content team. Whether or not you're fortunate enough to have competitive analysis reports to work from, it's well worth your team's time to better understand how the content you're creating holds up when compared to your chief competitor's.

Comparing Content Marketing from Company A and Company B

Figure 4.8 shows an example of a radar or spiderweb graph that compares content from two competitors—Company A and Company B. For this comparison, a sampling of content marketing assets was evaluated, including e-books and landing pages meant specifically to attract new customers. Company A's content was determined to be less customer friendly compared to Company B's content. In the chart, Company A's content scored only a "2.25" out of 5 for customer centricity, while Company B scored a "3." That's simportant for Company A's content teams to know—especially when they're conducting content research, as it flags a clear opportunity to improve content. For example, how might the content team dig in using research to better understand what exact words and phrases are making their customer experience not as customer friendly as it might be? What words and phrases and messaging could make their content feel warmer and more engaging to their audience, and therefore more competitive with Company B?

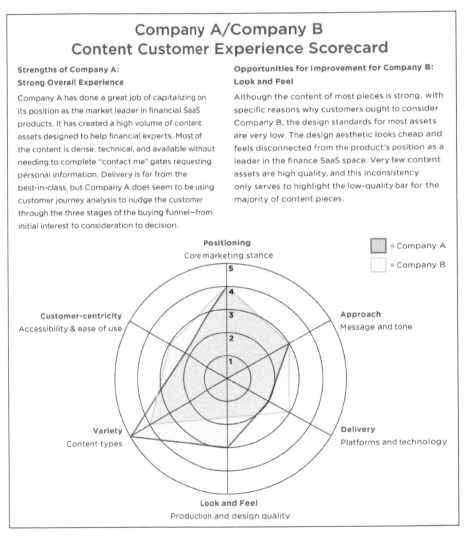

Company A/Company B
Content Customer Experience Scorecard

Strengths of Company A:
Strong Overall Experience

Company A has done a great job capitalizing on its position as the market leader in financial SaaS products. It has created a high volume of content assets designed to help financial experts. Most of the content is dense, technical, and available without needing to complete "contact me" gates requesting personal information. Delivery is far from the best-in-class, but Company A does seem to be using customer journey analysis to nudge the customer through the three stages of the buying funnel—from initial interest to consideration to decision.

Opportunities for Improvement for Company B:
Look and Feel

Although the content of most pieces is strong, with specific reasons why customers ought to consider Company B, the design standards for most assets are very low. The design aesthetic looks cheap and feels disconnected from the product's position as a leader in the finance SaaS space. Very few content assets are high quality, and this inconsistency only serves to highlight the low-quality bar for the majority of content pieces.

FIGURE 4.8

Radar (or "spiderweb") graph showing competitive analysis of content marketing assets for two competitors—Company A and Company B.

For each person outside of your content team who responds to your request for competitive analysis, keep a note. Add them to your list of stakeholders who can benefit from, participate in, and leverage the content research your content team runs. Be sure to keep in touch with each of them, to act as stakeholders or reviewers for your content research plans, to receive updates about the content research results, and to add to your recipient list for regular updates about content research results and impact.

Making It All Work for You

Evaluating the state of your content is time consuming, but well worth the effort. Before you eagerly start evaluating your content using content research, it's helpful to understand the lay of the land and how useful your content is overall—that is, how well it's reflecting your company's content standards and voice-and-tone guidelines, and where specific opportunities for improvement reside. Content evaluations require bravery, so be kind to yourself and your team as you take on this work.

CHAPTER 5

Identify Your Content Research Goals

One of the most frequently heard questions pertaining to content research is: "What content should I test?" Sometimes the answer to this question is obvious: You have a hugely important new feature launch in a month, and your team is focused on ensuring that the brand-new content for that experience is top-notch and as clear and engaging as possible.

Sometimes the answer is not so straightforward. Maybe you're on a content team responsible for all sorts of content—website, app, email, SMS, FAQ content, and more—and it's hard to know where to begin. Maybe you're a content team manager and your company leadership has asked you to "move the needle" and boost revenue by improving content overall and improving or "optimizing" the customer experience. That's a tall order. How do you go about tackling your complex (OK, if we're being honest, often chaotic!) content world and break it down into manageable research pieces?

Start with Your Most Important Content

If you've used the tools and tips in Chapter 4, "Evaluate the State of Your Content," then you've contemplated what is truly the *Most Important Content* (or *MIC*) for your team, company, or organization. You know and evangelize how your MIC is not necessarily the most popular or most visited content with tons of visits and click-throughs, but it's still integral to your customer experience. A terrific rule of thumb for content research is this: start with the Most Important Content.

Make sure that you think comprehensively and experientially about your MIC and how it truly behaves as part of the customer content journey in the real world, not in the idealized world of your product team. For the sake of example, if your MIC is a single landing page that promotes your company's most important product, break that landing page down into its content experience elements. *Each of these elements is a prime candidate for content research.* At first glance, your landing page likely has four or so obvious elements: the headline, subheadline, the body content, and a call-to-action (CTA) button or link, as seen in Figure 5.1. The CTA is always ripe for content research—never underestimate a CTA and its power to engage or to deter your audience!

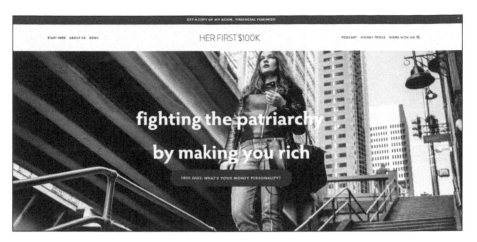

FIGURE 5.1

An illustration of a typical content landing page with specific content elements highlighted. Each element could be scrutinized by using content research to ensure that just the right words are being used.

However, you need to pause for a second. Yes, the landing page headline and its component parts are solid candidates for content research. But you also need to think about the landing page as part of the overall customer content experience. Just how *exactly* do your customers discover this landing page in the first place? What context or awareness of your product, if any, do they have at this point? (You may need to do some investigation to answer this question!)

There's another not-so-obvious content element that should be considered in your content research planning. It relates to how your customers find the landing page "in the wild." It's the SEO page title and the *meta description*, or the couple of short sentences describing the content that your customers see when they do a web search for your company or product. Surprise, surprise! So you don't have four potential pieces of content to research, after all, but rather six: the SEO page title, the meta description, plus the four parts of the landing page (headline, subheadline, body content, and the calls to action).

Google currently recommends that meta descriptions not exceed 160 characters, including spaces. While meta descriptions are no longer hugely important as SEO ranking factors—that is, in determining how high up in search results your content appears—they represent the "front door" of your content house, and therefore are essential to write clearly and descriptively in order to help your audience understand quickly and easily whether the content is going to help them accomplish whatever it is they're setting out to do.

A content ecosystem is not always what it seems at first glance. You can conduct research on any or all of these six content elements (see Figure 5.2). If you're tight on time and think you can research only one, then the SEO meta description is a great place to start, because that's quite likely your customer's first content experience impression—*not* the home page or landing page and its headline, subheadline, or body content.

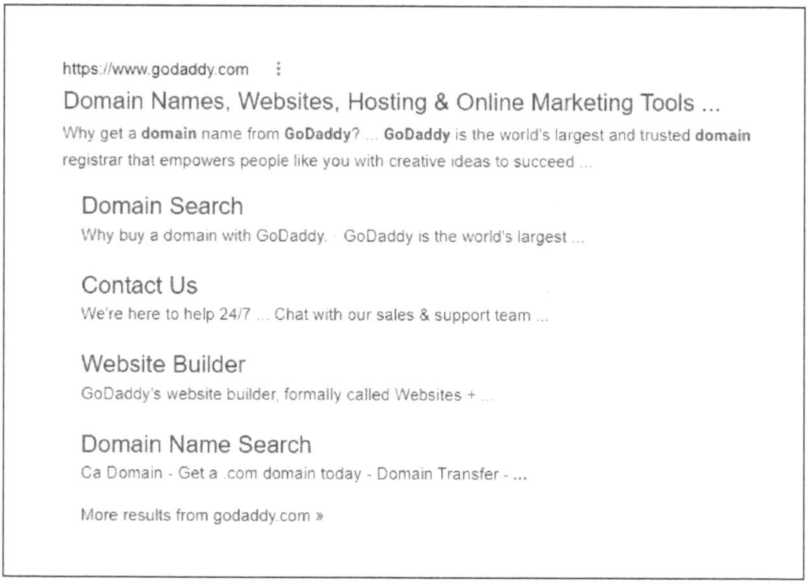

FIGURE 5.2

An example of a web search result, with a clear, well-written content page title and meta description, helps users quickly tell if the content associated with the description will help them accomplish their Job to Be Done.

If the meta description was never created, you're at the mercy of the search engines, because they'll "decide" what to display for the description. Their search bots will crawl through your content's code and then display either whatever content is encountered first, or whatever content their algorithms see fit. Not ideal! (To get a feel for what search engines see when they crawl your code, use the "View page source" or "Inspect code" commands from your web browser.) In other words, when you find a content page without a meta description, you've uncovered an example of "scary content"—content that's outdated, inaccurate, confusing to customers, or otherwise potentially harmful to your business and brand.

Some words of warning: Don't conduct content research on messy, outdated scary content! It's not worth your time and effort. Instead, update and refresh any scary content you encounter so that it's accurate, aligned with your terminology and style guidelines, and meets the quality standards outlined in your team's content quality heuristics. Content research isn't worth your time and resources unless you're conducting it on content that's already aligned to your content quality standards.

If the meta description does, in fact, exist, *and* it's aligned to your content quality standards, *and* you have plenty of search traffic to the landing page in question, then you can cross the meta description off your content research list. Leave it be. Move on to the next Most Important Content elements—the landing page headline or the call to action (CTA).

What If I Don't Have Time for a Content Evaluation?

If you're like many content strategists and content designers, you're extremely pressed for time and unable to devote resources to conduct a thorough content-quality evaluation, although you would really love to be able to carve out time to do so. (I see you, every one of you!) If that's the case, you can still use the tips here in this chapter to run content research studies. Sometimes, as challenging (and risky) as it is, you can only look forward and focus on the brand-new content your team is creating that hasn't yet been published or seen by your audience or customers and then make a plan to address the already existing customer-facing content at a later date.

Do keep in mind (always!) that your company's most frequently visited or used content is not necessarily its most important. A great recipe for research success is to keep your company or organization's top goals in mind, consider which core content experiences or pieces of content work to directly support those goals, and then conduct research on that core content.

Start Out Small and Scoped

When you're first starting out with content research, it's easy to want to run studies on *all* your content—as in, every single content element of your MIC, all the elements in your next Most Important Content, and so on!

Please fight the impulse to go wild with content research and rein in your coworkers if you notice them doing so.

It's best to keep the scope of your content research studies small and well-contained, with very specific goals. Keeping a narrow aperture on your research is important, at least until you gain stronger footing and build confidence in yourself as a content researcher, and you feel comfortable with the process. Fighting the urge to be incredibly ambitious will go a long way toward setting yourself up for success. Not only will it prevent you from feeling overwhelmed, but it will also make your research efforts easier on multiple levels.

The importance of scoping small:

1. When you have a small and focused study, it's easier to organize your research plan, and it's faster to gather input from your stakeholders and partners.

2. You'll be able to set up your research study more quickly and easily, no matter which tool or method you use.

3. If you include a focused, limited number of questions to ask your research participants—instead of a laundry list of questions—participants will be more likely to successfully answer all the questions you want to have answered. If you have an exhaustive list of questions included in a single research study, you'll find that participants get frustrated and stop providing as much detail in their answers—or they simply drop out and stop participating halfway through. (Have you ever regretted agreeing to take a survey where the questions just keep coming? You don't want the participants in your research to feel like that!)

4. With scoped, smaller studies, it's easier to communicate your insights and findings with your stakeholders and partners.

5. And finally, when you keep your research plans tight and succinct, it's easier to implement the insights you uncover from the research. Keep in mind that you may uncover a single "golden nugget" or content insight from a research study. However, if you're going to take that single insight and update your content to reflect the results of your research, that can translate into dozens or even hundreds of updates to your customer-facing content, depending on how much content you're responsible for. It's naturally easier to update your website or app to incorporate one or two key, amazing insights than it is to make a half dozen or more across your content ecosystem.

ONE CONTENT STUDY CAN MEAN HUNDREDS OF UPDATES

Keep in mind that a single content research question may result in your wanting or needing to make hundreds of updates to your team's content. Imagine that you're doing research for a national fitness studio franchise. Your app, website, and social media promotions describe your studio's workouts as "fun." From your research, you discover that most of your customers are enthusiastic about your gyms and describe the workouts as much more than that. The adjective that several research participants shared was "invigorating." That's a much more emphatic and memorable word compared to "fun." Using "invigorating" to describe the workouts could translate into stronger sales of gym memberships and keep current customers loyal.

You and your content team will probably want to jump on this newly found insight and act on it, to improve the content and customer experience. If you have an app, a website, a social media presence, emails, SMS messaging, sales team content (and so on) where you've been using "fun" to describe the workouts, then you'll have many instances where you will want to make the change to "invigorating." Of course, you don't need to go hog wild and immediately update every single instance of "fun." But you'll probably want to make that change at least on the most frequently visited or Most Important Content types or placements, and chip away at the other mentions as best you can over time, to make the messaging consistent across all touchpoints of the customer experience.

Remember, it's fine to pace yourself and your content team. You can always run a related follow-up study or multiple additional research studies to cover a single Most Important Content asset or experience. It's easy to feel sort of like a kid in a candy store when you first discover content research, and it's tempting to want to take advantage of research to evaluate all your new *and* existing content. Who doesn't want to discover insights about how well (or not) your content is resonating with readers, or find out more about your specific audiences so you can create stronger-performing content? Try not to let your research eyes be bigger than your stomach, and you'll be better able to use content research to your advantage and improve your user experiences.

Consider Your Content Research Altitude

Before getting started on any research, consider the altitude or level of the content that you'll be focusing on. Keep in mind all of the stakeholders who will be interested in your research and its results: C-suite and senior management, brand leaders, data analysts, marketers, product managers, software engineers, UX product designers, UX researchers, your social media team, and (of course) fellow content teammates.

Here are three main content research levels, listed from broadest to more precise:

- Product- or feature-focused content research
- Messaging framework–level content research
- Word- and phrase-specific content research

Product- or Feature-Focused Content Research

What do you know with certainty about what your audience wants or needs to know about your product, its features, or details? Are you providing enough information—or too much? Product- or feature-focused research is fascinating. It often comes into play on your home page, which means you'll need to collaborate closely with your marketing and brand teammates and other stakeholders to conduct this type of research. Marketing managers might want to list every feature and every detail—but are those relevant or overwhelming to your customers? You can do product-level research and find out.

Maybe your core product is a service or an app. How easy is it for your customers to use it? What additional info might they need to accomplish their Jobs to Be Done? In what content format would your customers prefer to receive additional information about how to use your app or service (video, animated tour, static content, or something else)? You can use content research to figure this out.

Messaging Framework–Level Content Research

Messaging frameworks (or content matrices) are documents that spell out how content for a specific customer experience varies, depending on any number of factors: the audience or persona, the content format, and the specific stage in the customer journey (see Figure 5.3). Messaging-level research identifies the most effective variant of content for each use case. For example, what CTA works best for Audience A (entrepreneurs) and which works best for Audience B (owners of medium-sized businesses)? Or what product description works most effectively on your website versus in social media posts? What product features do you need to tout to prospective customers at the very start of their customer journey versus when they're making the decision to buy your product or not? Messaging-level research can be especially useful for content marketing and when launching major (read: expensive!) promotional campaigns.

Messaging-level research is most helpful when conducted early on in your product and content development processes. It often yields fascinating insights about your customers' needs, thoughts, and motivations. If you and your team rely on customer personas, you're likely to find that messaging-level research can greatly inform those personas—*or even determine what personas you use or develop.* Sometimes, this research will utterly debunk the "facts" that are known by your teammates about your customers or audience, contradicting what you had previously believed was true.

The golden-nugget insights for messaging-level research are often found in the qualitative responses—that is, when you ask your customers *why* they prefer the words or descriptions or content format that they do, and why they don't prefer other choices. You'll get an earful and open your eyes to your customers' way of thinking. The result of messaging-level research is often priceless information and details that will help you connect to your audience better than you were able to before. And when you share what you learned from messaging-level research with your stakeholders, especially senior management, they will gain an appreciation for how critical content is to the customer experience.

Example of a Content Messaging Matrix

CONTENT ASSET	WHAT CUSTOMER PAIN POINTS OR CONCERNS THE CONTENT ADDRESSES	HOW CONTENT ADDRESSES CUSTOMER PAIN POINT OR "MOMENT OF TRUTH" (MOT)	SAMPLE QUOTE FROM CONTENT
Gated e-book "Amplify Your Sales Team's Productivity with Product XYZ"	• Concerns about total cost of ownership. • The need to provide quick return on investment.	• Drives home the immense productivity and efficiency benefits of using Product XYZ.	"Product XYZ can reduce administrative time by 15 to 20 percent, which means each seller gains an average of 53 minutes each day."
Blog post Company A named a leader in the latest Jones Research Report	• The need to drive revenue. • Why buy from Company A? (Social proof/trustworthy endorsement from respected industry analysis firm.)	• Highlights the Product XYZ as the highest scoring software product in its class. • Touts its machine learning, AI, and analytics capabilities. • Links to latest Jones Research analyst report.	"Company A delivers very strong analytics capabilities, including its machine learning and AI offerings."
Customer evidence profile (e-book) Interview/Q&A with CEO of Pike Place Technologies	• Change management: Need to know that software implementation will be smooth and quick.	• Profiles a booming business struggling under the weight of outdated technology. Tells how it supercharged its sales force to shorten the sales cycle. Lowered cost-per-call, and future-proofed its business.	"By using Product XYZ, we can supercharge and empower our sales team and shorten the sales cycle." –Jennifer Holland, CMO of Pike Place Technologies
Landing page Demo videos (5)	• Customers want to see solution in action before buying.	• Visually shows how Product XYZ provides a sophisticated, yet easy-to-use, data dashboard to boost sales team productivity.	"Real-time analytics based on historical data and predictive information provide sales leaders with powerful insights across their entire team."

FIGURE 5.3

An example of a content messaging matrix that outlines how content to promote enterprise software addresses specific customer concerns at specific stages of the customer journey. For example, the second content asset reveals how much the software costs, because that's a top question prospective customers have very early on in the decision-making process: "Can my company afford this product?" The last asset, which is emailed to customers as they get closer to making a purchasing decision, includes a video that demonstrates how the software works, answering the question, "How easy will it be to get my company up and running using this software, if we purchase it?"

Word- and Phrase-Specific Content Research

This is where you can settle many heated discussions among your colleagues, especially those outside of your content team!

What's *le mot juste* or the very best way to name or describe your product, features, or their capabilities, *according to your customers*? Discovering what words your customers use is integral to creating a solid customer experience. And content testing at the word or phrase level will help you nail this (and identify what words and terms your customers feel is jargon or is otherwise off-putting, overwhelming, or confusing to them).

Conducting content research at the word or phrase level also earns the rapt attention of everyone on your feature team. It's often fascinating, bias-busting work.

Which adjective is the absolute most preferred way to describe your core product? What word or words in the title of a content marketing video is most likely to hook in your audience? This is exactly when word- and phrase-level research is called for.

Word- and phrase-level research is probably also the easiest kind of research to do. Sometimes all it takes is two questions: a multiple-choice question to ask participants which word choice they prefer for the scenario (including an open-ended option for them to supply their own suggestion, if none provided fit the bill), followed by an open-ended, qualitative question that asks *why* they chose what they did (and perhaps why they don't prefer the other choices).

When you ask a sample group of customers—or people very similar to your customers—to tell you what words *they themselves* would use, you may as well be holding a pot of gold. When you know what exact words your customers prefer, and you work those words into your website, app, video, or whatever type of content you're creating, *it inevitably creates a more engaging customer experience*. Word-level research is a great way to prevent content "speedbumps"—those moments where customers scratch their heads when they encounter a word that's unclear or just doesn't feel right or ring true to them.

An Interview with Tracey Vantyghem,
Content Designer at Zipline

Tracey is a content designer at Zipline, an information technology company based in Montreal, Quebec,
Canada. She and her team ran a word-level research
study whose results were extremely surprising to
her and her colleagues in account management, front-end development,
marketing, and other departments.

As Tracey said, "I wanted to make sure that the words *our customers were
using* were the same that we were using in the content." One term that gave
her pause was "upperfield," which Zipline was using to describe people like
regional managers, who oversaw frontline workers in more than one location.

"One of the front-end engineers pointed out to me that we were spelling
it two ways: *upperfield* and *upper field*. I assumed it was industry jargon,
and therefore a prime example of an opportunity for content testing." She
added that her colleagues had "wildly differing opinions" about the term,
whether it ought to be replaced by something different, and if so, what
that might be. Tracey set up a simple content research study to understand
what word or term Zipline's customers would use to describe these frontline worker managers (see Figure 5.4). She sent the questions to several
dozen customers using the Intercom platform and received 77 responses.

Which word would you use to describe the group of people who support and oversee several *
field locations, such as all the locations in a region?

◯ Upperfield

◯ Upper field

◯ Field leaders

◯ Not sure

◯ Other...

Tell us why you chose this answer. If you chose "Other", why would you use this word instead? Walk
us through your thinking.

Long answer text

FIGURE 5.4

Tracey Vantyghem included questions in her content research study that
focused on the term "upperfield." The first question provided quantitative
insights, and the second provided qualitative ones.

continues

The team's research uncovered insights that were not only shocking, but a bit concerning. The net result was that content updates were needed to quickly fix a problem in the customer experience.

In other words, Tracey and her colleagues were blown away by what they learned. As it turned out, the word "upperfield" was perceived by Zipline's customers as worse than jargon. Only 18% of the respondents said they would use "upperfield" or "upper field" to describe the managers in question.

Of the remaining 72%, many felt the term "upperfield" was hierarchical and classist. Some were so confused that they thought "upperfield" was actually referring to *the computer screen itself.* In other words, they thought "upper-field" had something to do with when you fill out a form online—like filling out a form that asks for your name and address—and was referring to an empty field higher up toward the top of the computer screen.

When the team reviewed the results of this content test, "There was an onslaught of head-exploding emojis in our Slack messages!" she said.

After some further sleuthing, Tracey and her team found out that "upper-field" was a term that one of Zipline's first customers used—more than four years ago. The term unfortunately stuck. (Perhaps it stuck because the content design team wasn't staffed at that point and not involved in the decision to use "upperfield" on the website, in marketing materials, and so on.)

The majority of customers who responded to the study—57 percent—indicated that "field leaders" was clearer to them, as seen in Figure 5.5.

Tracey followed up the quantitative question with the qualitative question, "Please share your thoughts about why you replied the way you did." This qualitative question revealed that customers were quite confused about the term "upperfield," and that it contributed to deeper confusion about the overall user experience, which was ultimately undermining the company's brand and customer confidence (see Figure 5.6).

You can read between the lines and feel how flummoxed this person was by the term "upperfield." (The response rambled a bit and didn't always make sense grammatically, which is not unusual for content research responses.)

This person's confusion may have made it difficult for them to articulate their thoughts clearly and concisely.)

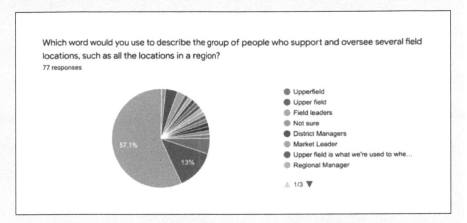

FIGURE 5.5
The pie-chart breakdown shows results from Tracey Vantyghem's content research study. After seeing these results, she and her teammates replaced "upperfield" with "field leaders" in hundreds of placements throughout Zipline's content.

> "I find Upper field confusing and when I did not know what it meant, "field" (as it relates to filling in content in a draft) refers to "the area where it is written" and "upper" makes me think of the location on the page since it's at the top. So I assumed it was "the top part of the message" and not "a group of people who are more senior"... Honestly, we don't use this and I don't fully understand the feature, so part of why I don't have a clear opinion... but if it was more clear, we might use it more!"

FIGURE 5.6
Here is a detailed response from one customer to the question, "Tell us why you chose the answer you did."

Which Research Question Type to Use and When

You're contemplating your Most Important Content and that content's altitude and stakeholders. Now, you can learn what question type to use when and view examples of each. The following isn't an exhaustive list of question types for content research, but it's one that's been used by dozens of colleagues and content-industry pros at many companies.

Here are seven basic types of content research questions:

- Clarity
- Comprehension
- Naming
- Preference (or Engagement)
- Audience-Specific
- Sentiment or Hedonic
- Actionability

In some instances, you'll see that there's a bit of overlap among the seven main content research approaches. For example, if you're interested in evaluating a call-to-action button, there is not one sole "right" research question type to use. (You can use clarity, comprehension, preference, sentiment, or audience-specific approaches; however, actionability is probably the most powerful approach for CTA buttons.) This overlap brings up an important point about content research: there is not one "right" way to do it!

Clarity Research

You can use clarity research when you want to know whether your users clearly understand a message, product or feature, or specific words or phrases. If the research shows that your audience doesn't understand the words you're using, you can use this research to identify the messages, product and feature names and descriptions, and words and phrases that are clearest and most engaging. Tracey Vantyghem's research shared in this chapter is an example of clarity research.

Which Content Types Are Ideal for Clarity Research?

Use clarity research to evaluate your customer experience when clearly understanding words or terms is key to customer success. That is, just about everywhere!

Navigation category labels are often referred to as *left-navigation* or *left-nav* labels for short, because they usually appear on the left-hand side of websites and apps. For example, if you visit an ecommerce site for a department store, and you're looking for denim, but you don't see a section in the navigation for "denim," you could easily think that the site doesn't sell what you're looking for. If the site is instead using the word "jeans" as its navigation label, you could easily overlook that section as well—and the website would miss out on making a sale.

Comprehension Research

Comprehension research is useful for when you want to find out how thoroughly your audience understands a message, feature or product, or word or phrase. Can your research participants accurately and thoroughly explain and define the term you're evaluating and describe it easily and confidently *in their own words*? Or do you find that most participants are providing halting, vague, or incomplete responses? (If the latter is the case, imagine if your question appeared on a high school pop quiz: your research participant would earn only partial credit!) Do you see evidence of your research participants pausing, hedging, and guessing?

When your customers can't for the life of them define a specific word that you use frequently in your content, this should set off your Spidey Sense. It indicates that your content is confusing or vague and incomplete, or in need of helpful definitions and explanations. What information or detail is missing? Simply asking your participants the qualitative question, "What, if anything, is unclear to you?" will help you learn exactly what's fuzzy.

Keep in mind that comprehension evaluations are often audience specific. Content that's clear and understandable to a resident of Miami may be confusing to someone in New York.

On the opposite end of the spectrum, you may learn from conducting comprehension research that your content provides *too* much detail. Again, your customers will let you know what the story is, and you can delete the extraneous information, or move it to where it truly belongs. For example, instead of including lots of details about a mobile app's customer interface, you may be able to simply move that information to the app's Help documentation or FAQs.

Comprehension research is excellent for navigation labels and for all sorts of other UX content, especially call-to-action (CTA) button labels and labels for tables and charts.

Comprehension research is also key to understanding gaps in your content. What information is missing that your customers really need to know?

Say, for example, you create content for a health insurance website. You want to make sure that your customers can understand terms like "copay," "deductible," and "coinsurance." To understand if your customers truly understand these terms, you can use content research and ask them to define the terms. Better yet, you can give them a sample scenario via content research. Ask your research participants a question that will reveal exactly what they understand and what they don't. Here's a great comprehension question to ask: "You have a health plan with a $2,000 annual deductible, and a 30% copay and $50 coinsurance per visit. Using the information and definitions provided here on this webpage, how much would your bill be for a broken ankle, if the cost of treatment is $750?"[1]

(This is sort of a trick question. At Premera Blue Cross, the digital experience team used this very question in content research. NOBODY got the answer right, including physicians and other people who worked in the medical field who participated in the research.)

Naming Research

Naming research is probably the most popular content research method. It has special usefulness in content marketing and in user experience content. If you find yourself having intense discussions with any of your colleagues (here's looking at you, marketers and product managers!) about what to name a new feature or a product, it's time to lean on naming research to help quell the back-and-forth and churn of subjective comments and opinions, and instead ask your customers what *they* think.

---

1   The answer depends on whether the annual deductible or annual copay maximum had been reached already or not. If none of the deductible had been paid, the patient would need to pay the full $750. If the deductible had been fully paid, then the patient would owe only the copay, or 30% of $750 ($225), plus the $50 coinsurance, for a total of $275. If the patient had paid their full deductible, as well as the maximum coinsurance, then they would owe only the copay, or $225.

Which Content Types Are Ideal for Naming Research?

Feature or product names are the most obvious type of content to be used for naming-focused research (see Figure 5.7). Naming-focused research is also helpful throughout UX content, for left navigation labels, pivot labels, and chart or table-column labels, etc.

Example of a Naming-Focused Research Study

Research Question: Imagine that you're an IT administrator of a large hospital chain. You're browsing on social media and see a promotion for an e-book. Which of the following names for that e-book would you find most appealing?

Response Options:

A. Reimagine Healthcare: Innovations That Are Saving Lives and Shaping the Future

B. Game-Changers: Technology Driving the Future of Healthcare

C. The Future of Healthcare Here Today: Key Innovations to Watch

D. Reimagine Healthcare: The Technology Shaping Our Present and Future

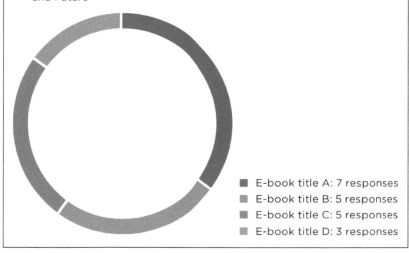

- E-book title A: 7 responses
- E-book title B: 5 responses
- E-book title C: 5 responses
- E-book title D: 3 responses

FIGURE 5.7

For this example of a naming-focused research study, a content marketer wanted to know which potential e-book title appealed most to the target audience. The e-book that included "saving lives" in its name appealed the most.

Preference (or Engagement) Research

Use preference research when you want to know which specific words or phrases your users like the most (see Figure 5.8). Preference research is sometimes referred to as *engagement research*, because it can often help indicate which words and phrases are most engaging to your audience. (There's logically some overlap here with naming research and actionability research.) The magic in preference research is often found from asking, "Tell me *why* you prefer the word you do." This approach often opens up a fascinating view into people's beliefs; reactions to voice, tone, and framing; and more. You'll also likely find that people tend to prefer words and phrases that are euphonic and soft sounding, rather than those that are cacophonic, with hard consonant sounds. How's that for some behavioral psychology overlapping with poetry and literary techniques?

Which Content Types Are Ideal for Preference Research?

Core words in your Most Important Content are ideal for preference research. You should evaluate the words in the headline of your most important landing page. Any part of speech is fair game. Ask your customers what nouns, verbs, or adjectives they like (and which they can't stand). You can use preference research to understand whether the verbs in your most important CTA buttons appeal to your customers, or whether they need more finessing.

Preference research is ideal to run prior to any A/B or multivariate content experiments that your team is planning to run. That way, you'll have pre-vetted your variable or experimental version of content and be able to "go out of the gate" that much stronger. This reduces the risk inherent in A/B and multivariate experimentation.

FIGURE 5.8

This multiple-choice question is an example of a preference question. (It also has elements of an actionability question.) In this example, two response options were preferred by eight and six participants respectively, while the other three options weren't chosen as often.

Audience-Specific Research

Use audience-specific, or targeted research, when you need to confirm what content is working for a certain audience. For example, if your company sells products for very small businesses and entrepreneurs, but you're starting to expand and sell to nonprofits, medium-sized businesses, or even large enterprises, you'll find some words are clear and appealing to some audiences, while they are confusing to another. You may have various audiences based on location, occupation or job role, company size, experience with your product or industry—you name it.

An example of audience-specific research is shown in Figure 5.9, where the results of two separate content research studies are placed side by side. In this example, two different audiences were asked how they would describe a data dashboard. Interestingly, the most popular response for both audiences was the same.

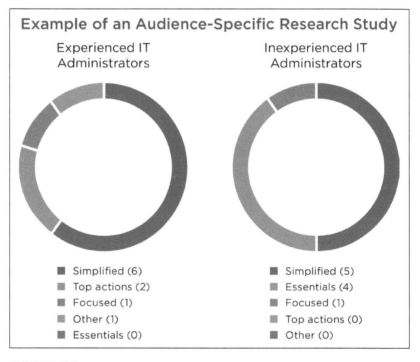

FIGURE 5.9
In this example, two research studies were run with different audiences—experienced and inexperienced information technology workers—asking them which name they would prefer for a new analytics dashboard. Stakeholders for this study hypothesized that more experienced IT administrators would balk at calling a dashboard "simplified." Interestingly, this study found that both experienced and inexperienced IT administrators preferred the "simplified" label.

Which Content Types are Ideal for Audience-Specific Research?

Audience-specific research overlaps a lot with clarity, preference, comprehension, and naming research.

Audience-specific research is also key for validating the accuracy of your localization and translation efforts. Is your content clear to customers in all the countries and regions you work with? Audience-specific research helps you know for sure.

Sentiment or Hedonic Research

Sentiment or hedonic research is helpful to understand how content makes a user feel. Do specific words or phrases lead users to feel more or less confident, pleased, unsure, happy, or unhappy?

You can also use sentiment research to find out how potential or current customers feel about your app, website, products, or services. Do they think they're reliable, and easy to use? Do they think they're innovative, or on the dull side? Do users think your products are "made for them," or do they feel unfamiliar and difficult to use, or somewhere in the middle of that spectrum? It's extremely helpful to find out why—especially if that "why" points to content as contributing to your user sentiment.

Sentiment research is an excellent type of research to combine with clarity research. It's also terrific to use in combination with heat maps and other click-tale type data. Your customers may be able to accomplish their Jobs to Be Done on your website or app, but how does doing so make them feel? Do they feel confident as they're setting up their new account with your company, or do they feel unsure and ambivalent? Asking your audience what top two or three adjectives they would use to describe their state of mind as they're using your content is an excellent way to determine if further work is needed on your word choices and voice and tone.

If you use System Usability Scale (SUS) or System Ease Score (SES) to evaluate your content, you'll have a few sentiment or hedonic measurements included in those scores (see Chapter 4).

Which Content Types Are Ideal for Sentiment Research?

Sentiment research is very helpful for marketing content, especially advertising or social media content, when you have a brief amount of time to capture your audience's attention. Sentiment research can also play a role in providing insights into (and sometimes debunking of) your content measurement and customer satisfaction metrics, such as Net Promoter Score (NPS). As mentioned earlier, Net Promoter Score is not a fantastic measurement to be using, especially for UX content. That said, if you can't escape the use of NPS by your team or company, you *can* use sentiment research to investigate whether your customers are feeling as displeased or pleased about your customer experience as NPS may be leading you to believe.

Actionability Research

Use actionability research when you want to know how likely participants say they are to take a specific action, based on the content they see. Actionability research can be especially helpful for campaigns and feature releases *prior to launch*. Are the words on a key landing page appealing or a turn-off?

Actionability research is also helpful if you're looking for ways to make very important content more engaging. The qualitative responses can be eye-opening (see Figure 5.10).

FIGURE 5.10

An example of an actionability research question and result. In this example, 12 people out of 20 said they felt they were "very likely" to share their email address after viewing the landing page that was shared with them in the research study. Most other respondents were also likely to share, with only one person replying that they felt they were "not at all likely" to share.

Which Content Types Are Ideal for Actionability Research?

Think about what customer actions are essential to your company or business and test the content related to those actions. If you send each customer a welcome email with next steps that they need to take, how likely are they to do those things? How soon after receiving the email do they think they'd take those steps (or not)?

(Keep in mind that what people say they will do is not always the same as what they'll actually do! Sometimes it's helpful to cross-check actionability research with content analytics, if they're available to you, such as heat maps and other "proof" of customer or audience behavior.)

Making It All Work for You

Conducting content research is exciting! That said, it's important to scope your work carefully, focusing first on the content that's the most important to you and your team. While it's tempting when you're first getting started with content research to run dozens of studies on as much content as possible, resist that urge. Identifying the Most Important Content is an exercise in stakeholder management and will likely require tough conversations. Do your best to refer to your content analytics and (diplomatically) help stakeholders who don't work on your content team to truly understand what specific content experiences drive the most impact for your business, and are therefore most worthy of content research.

Running research involves an investment of your time to set up, run, and evaluate the research, and—perhaps most importantly—to create a report of your findings. To keep your content research efforts organized, be sure not to bite off more than you can chew. You can always create a "backlog" or running list of prioritized research you and your team want to run in the future.

Understanding the core types of content research—clarity, comprehension, naming, preference, audience-specific, sentiment, and actionability—is key to helping you and your team scope your studies. (Sometimes, research will overlap and cover two or more of these categories.) Stay laser-focused on what you truly need or want to learn about your audience and keep your research studies short, clear, and participant-friendly. You'll feel your research program take off, generating positive energy and uncovering insights that will benefit your whole organization for months and even years to come.

CHAPTER 6

Research Planning and Stakeholder Management

A fter you've determined what specifically you need to learn about your content, here comes the exciting part: planning, structuring, and running your study. When it comes to content research, success lies in thorough preparation—just like painting a room. It may feel clunky and time-consuming the first few times you prepare for a research study. However, when you keep at it, it gets much easier and faster.

Balance Who Is Involved in the Research

One of the most common mistakes made by those new to content research is allowing too many cooks in the kitchen. It's exciting to see the energy that content research creates, and that means people will hear about and ask to be part of the work. But allowing too many people to participate in the key decision-making part of the content research process will slow you down. Here are a few things to keep in mind:

- **Don't fly solo.** It's so fast and easy to set up a research study using online platforms, like UserTesting or dscout, that you may be tempted to conduct research as a lone wolf, without first obtaining stakeholder input. This is not recommended! Omitting key stakeholders is not a good idea—ever. This is especially true when you go to share research results. If the research results conflict with your stakeholders' hypotheses or expectations, you'll hear about it. It's simply best to gain stakeholder input up front.

- **Think about post-research implementation.** When your study is done and you have insights ready to be implemented into your content, you may need help from your colleagues to do so. It's awkward to ask a product manager or other coworker for help making content updates if they weren't aware that the research was happening in the first place.

- **Avoid rework.** No matter who you involve in research, you'll inevitably get requests from other people who want to add their two cents' worth to the design of the research study or add a key question they're eager to get answered. If you establish clear boundaries to your department ahead of starting the research study as to who is accountable and informed about the research, you can help avoid this type of rework.

The safer, smarter approach is to keep your short list of stakeholders informed of and involved with your research plans, and aware of when a study is in progress. That way, they won't be surprised when you share the study results. When it's appropriate, bring a few (say it again: a *few*!) select stakeholders directly into the fold and involve them in the research planning itself. Read on to find out how to do just that.

Consider a Responsibility Framework

Word sure travels fast when you run content research. This is great, since you want lots of people in your company to understand how important content research is—and to be enthusiastic about it! However, for the purposes of getting each study completed and its findings implemented quickly and easily, you absolutely need to limit the number of people involved.

If you find the number of colleagues who *want* to be or say they *need* to be involved is becoming unwieldy, you can turn to a project-planning tool called a *RACI chart*, which is a simple one-page chart that lists the names and roles of the key people involved in a project: *responsible, accountable, consulted,* or *informed.* Figure 6.1 shows a sample research plan template.

This kind of framework will help you accomplish many things:

- Establish who's expected to do what. Role clarity is key!
- Prevent hard feelings or misunderstandings from those who are not involved in all the decision-making.
- Obtain input from only the colleagues you truly need to have involved, and not get slowed down or overwhelmed by input from others.
- Build trust among your colleagues because you're making it clear that you'll keep everyone informed at key touchpoints during the content research process.
- Work more quickly, as defined roles and clear responsibilities lessen the chance that someone will "step out of their swim lane" and step on someone else's toes. Defined roles also support accountability, so people are very clear about what's expected of them.

Table 6.1 shows a sample role-and-responsibility chart that's useful for content research. As the content lead, you are the responsible one. The other roles will differ based on your company structure, industry, product, and so on. If you have any doubts about whom to include, include a representative from each of the key business functions you work with on a regular basis and ask them to be sure to communicate all project updates with the appropriate people on their team.

TABLE 6.1 SAMPLE RACI TABLE

Responsible	Accountable	Consulted	Informed
You!	Product manager lead Product design lead Others as needed	Legal lead Marketing lead Software development lead Others as needed	Your manager The managers of other "accountable" colleagues Others as needed

Keys to Success with Stakeholder Frameworks

The key to being successful when using a responsibility framework is *not* to let people identify and claim their own roles. *Au contraire!* What you need to do as the content research lead is to take charge and determine, as best you can, the short list of who truly needs to participate, and at what level of involvement. That means clearly calling the shots and establishing who will be fully involved and need to sign off on the research plan, as opposed to who will participate in the research planning and execution, or who will be (politely) informed but won't directly participate. Then you need to clearly communicate that information to your core stakeholders. (It also means that you need to communicate with the stakeholders who aren't on your short list and gently ensure that they're clear about how they are *not* expected to participate.)

Yes, this will mean that you may be the bearer of somewhat disappointing news—for example, in the case where you need to let someone know they're going to be *informed* only, but they want to be *accountable* as well. Make sure that you're careful and sensitive when informing your stakeholders of their level of participation.

If you work for a large company, or you work in a field that's highly regulated, like finance, healthcare, or education, your legal or policy team may need to be involved and provide a review and approval of your test plan before you run the study. When in doubt, ask and make sure. You don't want to waste your time or that of your coworkers by conducting research and having your legal team temporarily or permanently block the research insights from seeing the light of day because the research didn't align with company policy. (See more on legal considerations later in this chapter in "Company Policies, Legal Requirements, and NDAs.")

Outlining the Test Plan

After your responsibility framework is set, step 1 is getting up and running with creating the test plan. To keep this process as light and quick as possible, it's a great idea to invest some time into developing a content research study template, which can be as simple as a quick one-page outline of the most important information related to your research.

Here are elements of a typical, sample content research plan:

1. **Content Research Goal:** What specific, scoped insights are you aiming to learn and for what specific message, product, or feature? (See Chapter 5, "Identify Your Content Research Goals.")

   A key element to keep in mind as you set your research goal is your business scenario. What does success look like to your product manager, to your manager, and to the digital team overall? To the best of your ability, ensure that the content research goal supports those organizational definitions of success.

   Remember to keep the research goal focused and specific. This pertains to both the length and breadth of the content you're evaluating, and for the audience. Aim to have your research study require a maximum of 15 or 20 minutes of your participants' time, especially if they'll be providing feedback online instead of in person. Otherwise, attention spans can lag, and you may get incomplete or muddy input from participants impatiently speeding through their responses.

2. **Sample Goal:** Your goals will directly determine the format of the research questions you'll need. For more information on creating your research questions, see Chapter 7, "Craft Your Content Research Questions."

   Here's a potential sample research goal: *What specific adjectives or descriptions of Product XYZ do entrepreneurs find the most—and the least—appealing?*

3. **Nongoals:** What is out of scope or *not* a goal of your study? What are you deliberately *not* going to address with your research, to keep the research study focused and tight?

   (Remember that scoping your research is key to being successful!) For example, if you're focusing on an audience of entrepreneurs who work at companies with five or fewer employees, it's worth stating that business owners with more than six employees are not going to be included in the study.

   This may seem like overkill, but bluntly including the details about what's out of scope will help you and your stakeholders to progress efficiently and will also support you in redirecting any conversations that start to veer into out-of-scope territory so that you stay focused on the scoped plan.

4. **Scenario:** Briefly describe which specific content experience you're investigating, along with a short description of the content experience. For example, are you curious about prospective customers as they visit Product XYZ's landing page and consider purchasing it for the first time? Be sure to specify what product or feature you're evaluating, and the specific, scoped content element(s) related to it.

5. **Key Stakeholders:** List which colleagues are part of the core team involved in this study and their email or contact info. Include yourself (or the name of the content lead) and ideally a single representative from the other teams involved in your core, day-to-day content development process. For a typical product team, this would include a product designer, product manager, and software developer, and possibly a marketer and a data analyst.

6. **Methodology Tasks and Questions:** This section includes a short list of specific questions you're going to be asking and brief details about the format you're using for the questions (which speaks to your methodology). It also includes which tasks, if any, you are requesting that participants perform. This usually involves including screenshots of content for participants to evaluate, or links to interactive prototypes for them to navigate through or review before answering questions. You'll find more approaches for methodologies and question formats in Chapter 7.

7. **Audience Details:** This refers to the qualities you are requiring of research participants, such as their location, company size, or other specifics integral to helping you accomplish your research goal. Note that there is a bias inherently involved in audience targeting. Tread carefully and sensitively! (There's more information on bias later in this chapter, including steps you can take to help reduce bias and resources for additional learning on this topic, as it is extremely important.)

Don't Test What You Can't Implement

This concept is mentioned in other chapters of this book and bears repeating: set yourself up for success. Be careful when conducting content research—especially when you're using questions in a multiple-choice format—so that you're not inadvertently asking your participants to weigh in on any content options that you and your team shouldn't or truly cannot implement. There can be many and multiple reasons why you shouldn't or can't publish certain content, including branding, legal, competitive, sensitivity, or other reasons.

Beyond legal and branding reasons, one key thing to be on the watch for with content testing is basic, but so important: the length of content options you test. Make sure that you and your stakeholders are clear on any content length requirements (such as if you're testing social-media headlines or names of content assets that may have a maximum length due to your content management system, marketing operations platforms, or other tools). If you and your stakeholders are not on the same page about maximum content length, you may find that the "winner" of a content study is unusable, and that will sadly render your content research efforts to be a waste of time and resources.

Product XYZ Naming Study

Type of Content Study: [Example: Unmoderated, 10 participants from very small businesses]

Tool: [Example: UserZoom.com]

Stakeholders: [names & email—include product management, user research (if needed), product design]

GOALS:

1. What business/customer questions are you trying to answer?
2. What do you hope to learn?

NON-GOALS:

What questions are you NOT answering about this experience (for example, related questions that have already been answered in previous studies)?

SCENARIO:

[Example: You have a small business with just you and five employees, and you just signed up to use Product XYZ for the first time. After purchasing it, you are taken to this webpage.]

TASKS FOR THE CONTENT RESEARCH STUDY:

Participant Instructions:

1. Select the URL to go to the prototype (or view the screenshot).
   [Add Prototype URL here—for example, link to Figma screen.]
2. Please answer these questions about the information shown:
   [Examples]:
   A. What words would you use to describe this page? [Open-ended question format]
   B. When you look at this page, how does it make you feel about managing Product XYZ for people in your company? (Scale/Likert question, ranging from very anxious [1] to very confident [9].)
   C. Which of these statements seems most appropriate to describe what is shown?
      * It's a dashboard.
      * It's a home page.
      * It's a specific view of the home page.
      * It's a mode for viewing the admin center.
      * Other
   D. Tell us why you chose this answer. If you chose "Other," what would you say instead and why? Walk us through your thinking.

AUDIENCE TARGETING:

Summary of who the target audience is for the test.

1. Company size (very small, small, medium, large, enterprise)
2. Countries
3. Income range
4. Employment status
5. Other screener questions to help ensure that the test is sent to people in the audience needed

FILTERS:

☐ Age	☐ Social networks	☐ Household income ($)	☐ Other requirements
☐ Gender	☐ Parental status	☐ Job level	☐ Web browsers
☐ Industry	☐ Operating systems	☐ Web expertise	(Defaults to Chrome)
☐ Job role	☐ Test participation frequency	☐ Language	☐ Participation in prior studies

Ex: To target to users of Product XYZ, you can ask:
Are you a regular user of the Product XYZ?
Yes (accept)
No (reject)

Ex: To target to very small businesses, you can ask:
How many people, including yourself, work for your company?
1-9 (accept)
10-99 (reject)
100-399 (reject)
400+ (reject)

FIGURE 6.1

A sample research study template that you can use to help create your own research studies.

Organize a Stakeholder Working Session

Working sessions are collaborative meetings where work gets accomplished quickly, because the right people are present and able to immediately answer any questions that come up on the fly. Collaborative working sessions call to mind the expression "many hands make light work." Drafting your content research plan is a perfect opportunity for a collaborative working meeting or session with the core stakeholders on your content research A Team. (If you're using a RACI chart, that means you include only the few special people who are *accountable*.)

Say you have a content research study template and a handful of carefully chosen colleagues whom you've identified as your right-hand partners in content research. Fire up a Google Doc or Word document to create the draft of your content research study plan, making sure that it's shareable and editable by everyone on the A Team. Schedule a half-hour-long working session with your group and hammer out the details of the research study together.

As the project lead for your research efforts, you may already have your research template partly or mostly filled in ahead of this meeting. If not, fill out the research template as you and the A Team meet and discuss the plan, collaboratively and quickly updating and editing each item in the template, making sure that everyone's voice is heard and that everyone is on the same page.

During the working session, get the key details of your research study plan captured as cleanly and succinctly as you can. Remember, by documenting the research study plan, you're accomplishing many things simultaneously. You're preventing future surprises among your stakeholders. You're laying the groundwork to get the content results documented easily, to reflect what your team learns from the research. You're also setting the stage to get team support for updating your customer-facing content to reflect the golden nuggets of content insights from your research. Depending on how you update content for your company or organization, those nuggets may require time and work from your product design peers, product manager, and software developer counterparts. You also may need to share your documentation with your data analysis team, if they need to adjust their content tagging, if and when you implement the golden nuggets that were uncovered during research.

Include a Visual of the Content Under Evaluation

If you're researching content that's already been published (also called *customer-facing content*), it's handy to include a screenshot or image of the content that's being evaluated in the research-plan document. That way, everyone has a clear visual and is familiar with the content in question. Since not all of your stakeholders likely have access to your design files, content management system, and so on, including a visual or screenshot that gets all stakeholders on the same page is a great way to make sure that you get alignment on the work at hand. You may also find that the jargon or terminology you use as a content professional ironically may not be the same as what your product manager or other stakeholders use. A "landing page" to you or "app start screen" may not be immediately clear or may mean very different things to others involved in the research.

Making sure that there's a visual reference for you and your stakeholders is also key because chances are that different stakeholders will have different recollections or images in their mind about the content in question. Are you and your team researching the desktop experience, the app, or the mobile version? By providing a frame of reference, you can make sure that your core stakeholders are all on the same page so that the research progresses smoothly.

Identify Your Research Platform and Approach

There are dozens of ways to conduct content research. It can be conducted in person—when it's safe to do so—or online. The four main types of research approaches are:

- Unmoderated online research
- Moderated online research
- Moderated in-person research
- Person-on-the-street interviews (sometimes unfortunately referred to as *guerrilla research*)

When research is *unmoderated,* what that means simply is that you don't need to be present. Usually, this means that research participants are asked a fixed list of straightforward questions. Online surveys are a typical example of unmoderated research.

When research is *moderated,* it means that you or someone on your team must be present while participants are answering the research questions. Moderated research is the most complicated to conduct and analyze, since you can adjust, adapt, change, or add questions on the fly, depending on the responses you're hearing. For example, if a participant mentions that a specific word is completely unclear (or offensive or otherwise bothersome) to them, and you had no inkling that this word might be unclear, that's an opportunity to dig deeper and ask for further details. These on-the-fly adjustments can also mean that it's tougher to analyze the responses.

PRO TIP WHY NO FOCUS GROUPS?

Focus groups are conspicuously and deliberately omitted from this list of research approaches. Why? First, in-person research with individuals is not always affordable or feasible. Second, it's awfully easy for people in focus groups to have their thoughts and opinions (and therefore their statements and comments) swayed and influenced by others present in the group. Third, focus groups can be intimidating and stressful for introverts, and conversations can easily become dominated by the extroverts or most outspoken people in the group.

While it's more time-consuming to conduct research with individuals, you won't run the chance of having groupthink potentially influence or skew the feedback received.

Research Approaches Range from Simple to Sophisticated

Research approaches can be as simple as using email to send questions to your research participants or phone calls made to participants where you ask them your research questions. You can also conduct spontaneous on-the-street interviews. Tools can vary from good old pen and paper to Google Docs or Microsoft Word, to sophisticated online platforms that are continually adding new capabilities and features, like UserZoom, dscout, Optimizely, or Qualtrics, or an online survey tool like Momentive (formerly known as *SurveyMonkey*). The approach and tools you use will depend on your timeframe, your team or company budget, and your ability to influence those in your organization (to provide a budget for the tools, if needed).

The annual cost of these online research platforms varies and often depends on the size or revenue of your company. For an enterprise-size (global) company, an individual license for one of these platforms can run more than $4,000. That said, these tools can literally—and that's meant in the true sense of the word!—pay for themselves by helping you improve your content so that it attracts new customers, more effectively engages current customers, and saves your company significant money by preventing customer-service calls, chats, and emails. Before it merged with UserZoom, UserTesting cited a statistic from Forrester Research, a leading technology market research firm, which stated that the return on investment or *ROI* for usability research was at least $2 for every $1 spent on research and as high as $100![1]

> **PRO TIP** PARTNERING WITH STAKEHOLDERS TO FUND RESEARCH
>
> If your content team currently does not have the budget to pay for an online research platform, it's worth seeing if you can part-ner with your colleagues in marketing, engineering, or product management to see if they do. Since the entire product team will benefit from clearer content, it's worth reaching out and asking.

UserZoom, dscout, and Other Online Research Tools

Online user research tools and platforms like UserZoom, dscout, Qualtrics, Momentum, and so on are immensely helpful. Each has online guidelines, as well as tips and tutorials to help you learn how to use them quickly. Keep in mind that many of them were created for capturing feedback on user experience designs, or prototypes, and *not* on content or content design specifically. The information in this chapter and in Chapter 7 will help you leverage the platforms to use them successfully for content-focused research purposes. Figures 6.2 through 6.4 show the UserTesting, UserZoom, and dscout home pages.

---

1   UserTesting, "Proving the ROI of UX Research," https://info.usertesting.com/rs/220-GOX-255/images/Proving_the_ROI_of_UX_research.pdf

FIGURE 6.2

The UserTesting home page includes a link to its extensive resources section and a sample video from a user research study. UserTesting also offers a UserTesting University, a collection of articles and video tutorials to help you get up to speed.

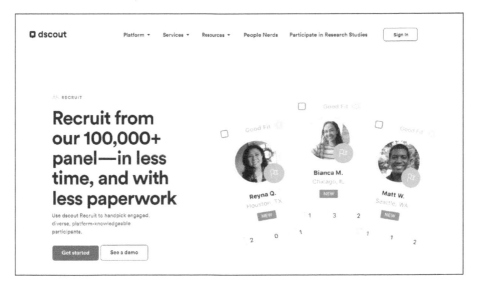

FIGURE 6.3

The dscout "Recruit" landing page mentions that the platform focuses on diversity and has more than 100,000 people ready to participate in research.

FIGURE 6.4

The UserZoom home page displays the many qualitative data formats available from a typical research study.

Finding Participants Without Help from an Online Research Platform

If you're conducting in-person or online research without using an online research platform, you'll need to find research study participants on your own. This isn't that difficult if you have a customer service team you collaborate with. That said, before you proceed any further, *make absolutely sure that it's acceptable from a legal and company policy point of view to contact your customers!*

Company Policies, Legal Requirements, and NDAs

Whether you lean on an online platform to recruit research participants for you, or you tap your customer service team to be put in touch with customers, make sure that you're working within your company's and industry's policies and regulations. If this sounds a bit scary, it should! If you work in a regulated industry like finance or healthcare, or in a country with stringent consumer data privacy regulations (such as Germany), make sure that you're not conducting research that's in violation of any of those rules or regulations. *When in doubt, contact your company's legal team or Chief Operating Officer to make sure that your research plans won't be getting you or your company into any potential legal trouble.* **Do not skip this step!**

In addition to terms and conditions, you need to be mindful of PII, or personally identifiable information. Conducting research puts you at risk for obtaining PII, which may violate your company's policies. To address this issue, often researchers will add a bold disclaimer at the start of their research study: "Do not share personally identifiable information with us when you answer our questions." However, there's no guarantee that participants will notice or follow that request. (You'll probably also need to define PII, as it's not a household term.) Err on the side of caution.

If you work for a startup, or you're researching content for a brand-new feature or product for any size company, you'll likely want or need to ask participants to sign a nondisclosure agreement (NDA). Make sure to keep all signed NDAs in a safe place for future reference.

Finding Current Customers to Participate in Your Research

Let's assume that you received the all-clear from your legal team to reach out to a few dozen customers to conduct research. One of the simplest ways to find research participants is through your customer service or customer experience team. Make friends with your customer service leads, because they'll be pivotal in helping you obtain customer contact information. If your company uses a customer relationship management tool or platform like Zendesk, your work is made that much easier.

Do your best to reduce bias in how you select customers to participate in research. If you have a database or list of customers who may be safely contacted, try to select customers as randomly as possible. If you can mask or "de-identify" their names, to help reduce bias, do so. You can also use a website like Random.org to "assign" a number for each customer name, which can help you select as random a sample of people as possible to participate in your study.

Do keep careful track of the customers you contact to participate in research, so that you don't reach out to a customer more than once. A good guideline is to contact customers directly only once a year.

Finding Participants Who Are *Not* Current Customers

What about when you want a completely fresh perspective from your research participants? That is, what if you do not want to conduct research using current customers and instead want to find people who are potential customers, or are "lapsed" or former customers? In this situation, you may want to employ the use of an agency or service like Ethnio or User Interviews, whose specific purpose is to help find and recruit people to participate in surveys, usability studies, and so on. Doing so will add to the cost of your research. However, like with all customer research, the expense will likely make up for itself and more in improved content quality, which translates to better customer engagement, revenue, and so on. Figures 6.5 and 6.6 show images from two websites that help connect researchers to participants for UX research studies, Ethnio and UserInterviews.

FIGURE 6.5
The home page of Ethnio, which is a UX research participant recruitment site.

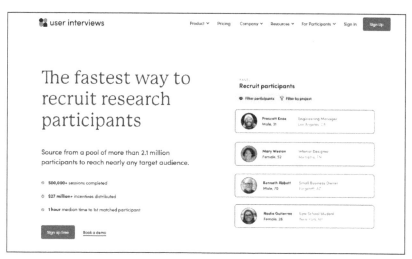

FIGURE 6.6
The "Recruit" landing page of User Interviews, which helps UX researchers find participants for online UX research.

Being Respectful About Online Study Participants

An important plea: Please be mindful about the need to be sensitive and be respectful of your research participants. Even though these individuals are often anonymous, or you know them only from their online usernames, you will be able to obtain audio and video recordings of participants as they complete your research studies. Because you'll sometimes be aware of their age, income range, and other personal characteristics and qualities, it's important to be sensitive as you create, run, and analyze your research results. *You're working with real people who have real feelings and who deserve respect and consideration.*

Sometimes, you'll observe that online research participants may choose online usernames that may appear a bit playful, or even veer into what you might consider silly. Don't assume that just because someone is using what seems like a quirky online name (for the sake of example, "EspressoLover" or "ManicPanicGuy") that they don't have valuable contributions to offer your research program.

Choose the Number of Participants

One element of these online research platforms that's often surprising to people who are new to online content research is that they limit the number of test participants. *This is often a good thing!* Ninety-nine percent of the time, you don't want to run a study with 100 or 1,000 participants. If you run a huge study, that means that when you get the study results, you'll be spending hours and hours wading through the responses, especially if you include qualitative or "why" questions. And you *do* want to include qualitative questions in your studies whenever you can, as those are where the "golden nugget insights" are often found.

It's perfectly fine and, indeed, preferable to run a content research study with just a small handful of participants. Five, 10, or 15 participants is often perfectly sufficient for getting directional feedback on your content that you can use to make positive improvements in your content. And remember, you're working on content, not prescription drug development; you're not rigidly seeking statistical significance, but rather just enough feedback to help you identify insights and patterns and ensure that your content reflects the thoughts and preferences of your target audience.

If you choose to run a study with a small, even number, such as 10 or 20 participants, you'll be making life easier for yourself when you need to share quantitative results. When you have an even number to work with, it's fast and easy to take the responses to any questions that are in a multiple-choice or scale format and multiply by the appropriate number to create percentages. These percentages are key when you're working with stakeholders and reporting your results to other senior leaders (such as in monthly or quarterly business reviews).

Specify Your Participant Criteria

The most convenient and valuable part of online research platforms is that they will share your online test or study with specific people who fit the profile that you specify: location, age, gender, language spoken, job title and level of seniority, company size, the industry they work in, income, and so on.

You can also add custom "screener" questions to further refine the audience to suit your specific needs. For example, UserTesting's test template offers three ranges of company size, ranging from 1–50, 51–1,000, or over 1,000 people (see Figure 6.7). These ranges are fairly standard. However, if your company's or team's content or marketing personas and goals don't align to those default ranges, you'll need to add a screener question or two to help make sure that your test platform sources participants as closely as possible to the specific audience you're after.

For example, say you're doing research specifically on "solopreneurs," so you're looking for people whose company size is just 1. Or maybe your marketing team's persona of "medium-size" businesses is defined as companies of up to 500 employees. Neither of these cases will be served by the platform's default ranges. You're going to need to create some custom screener questions. This can be done rapidly. (Sometimes stakeholders grow concerned when a research platform's default criteria don't perfectly match what you need. Reassure them that this is nothing to worry about!) Figure 6.7 below shows a sample default company-size filter.

FIGURE 6.7
A company-size filter, like this one from UserTesting, may not be flexible enough to suit your testing needs. In that case, you'll need to add additional screener questions.

Filters

Company size

☐ Small 1-50
☐ Midsize 51-1000
☐ Large 1001+

Figure 6.8 shows an example of additional screener questions, to help target a study to *very small* business owners of 10 or fewer employees.

What size company do you work for?
- ☐ I'm a solopreneur (I'm the only person working at my company)
- ☐ 2-10 people
- ☐ 11-50 people
- ☐ 51-99 people
- ☐ 100-499 people
- ☐ 500-1,000 people
- ☐ Over 1,000 people

FIGURE 6.8

These screener questions are intended to identify research participants who work in small companies with 10 employees or fewer. If a respondent answers "yes" to the first or second question, they will be allowed to participate. If they answer "yes" to any of the last five categories, they will not.

Understanding Online Research Participant Pools

Determining the specific participants whom you want to reach is probably the most important step when you prepare to run your study. That said, to find the right participants, you need to understand how the online platforms source their participants.

Online platforms like UserZoom and dscout have thousands of people around the United States and in some cases, around the world, who are registered in their "pool" of research participants. These individuals are compensated for each research study they participate in. When they sign up to participate in research, they provide lots of personal information, including their age, where they live, whether they're employed (and if so, where they're employed, their industry, level of seniority, years of experience, and income range).

This information is personal and private and is also the information that helps your research studies get shared with the "right" people—the people who reflect the specific criteria you're looking for. Maybe for your research purposes, you're looking for women over the age of 50 who live in the U.S. and use iPhones. Perhaps you're looking for people between the ages of 40 and 60 who work in senior-level roles in the hospitality industry in Europe. Online research platforms can help you find people who have the specific criteria you're looking for, so you get valuable, actionable feedback on your content experiences (and save hours of time by not having to search for these people yourself).

Balancing Specificity and Speed

When conducting research, you, of course, want to find participants who are as similar as possible to your "target" audience. And one of the most incredible things about online research is just how quickly these online platforms can find participants and get your study completed in a short period of time—sometimes as quickly as an hour, if you are looking for basic criteria.

However, if you have a niche or very specific audience in mind, it can take days, or sometimes weeks for enough people who fit the bill to be identified by your testing platform and sent your list of questions. In some cases, your research may become stalled because the platforms simply can't find the people you're looking for.

On the other hand, you don't want to be too general and broad when identifying the criteria for your preferred research participants, or else you may find that the people who participate in your test aren't quite the respondents you want. If you use a too-general audience, you'll have a harder time finding insights among the responses and the research results won't be as actionable as they might be. You may find you even need to redo your study with more specific targeting.

While online research platforms usually do a pretty good job of finding people to participate whose characteristics align with the criteria you specify, please be aware they're not foolproof. It's a balancing act to identify the right participant criteria, and it takes some trial and error and practice to refine your audience-targeting technique.

The Conflict About Compensation

Remember that your study participants are compensated for their participation; they don't answer your questions for free! Keep this in mind, as sometimes people show a tendency to be overly solicitous or complimentary about your content, because they feel compelled to say "nice" things simply because they're being paid. As you gain experience running research studies, your Spidey Sense will grow sharp, and you'll be able to identify test participants who are far too polite, and whose feedback may result in "outlier" results that aren't as valuable to your research as other participants.

To help head off this phenomenon at the pass, you can add a quick bit of candid guidance at the very start of your study. Encourage participants to be honest about their feelings and reactions to your content. Remind them that their constructive feedback is exactly what you're looking for and precisely what will help you and your company improve your products or services.

(In your case, the *product* you're improving is *content*—but the research participants don't need to know that and, in fact, shouldn't know it, or it may overly influence their responses.)

REDUCING BIAS WHEN SEEKING CONTENT RESEARCH PARTICIPANTS

Note that this section is called "reducing" bias, not "removing" bias. It's simply not possible to omit all bias from content research, and especially when using online platforms—although that doesn't mean you shouldn't try. It's important to be sensitive and mindful of diversity and inclusion when creating your study plan, specifying your preferred participants, and analyzing and sharing your study results.

For example, online research platforms allow you to specify the income and employment status of participants. Your stakeholders may want to include only people who are currently employed in a specific industry that's relevant to your product, feature, or messaging research. Please be open-minded and encourage open-mindedness among your stakeholders. Simply because someone identifies as unemployed, that doesn't mean they don't have valuable opinions and information to share. Perhaps they have recently retired. Perhaps they're on family leave or sabbatical. It's worth expanding the aperture of your research targeting to be open-minded and inclusive, and to encourage your core group of stakeholders to be open-minded as well.

At the time of this writing, online research tools were not inclusive of gender identities, with many requiring participants to identify as either male or female, or to decline to provide a response.

continues

Another very serious issue with these tools is that, currently, it's impossible to know if there's racial diversity among online participants. These are serious weaknesses, and it's worth recognizing and being continually mindful of during your research work. It's also important to call out this fact when you write your research report and share results with your organization. Addressing bias in user experience is worthy of a book of its own. David Dylan Thomas's *Design for Cognitive Bias*[2] is an excellent reference.

Figure 6.9 shows a collection of types of cognitive bias, compiled by Buster Benson, a former product manager at Twitter, and visualized by John Manoogian III, a product manager at Apple.[3]

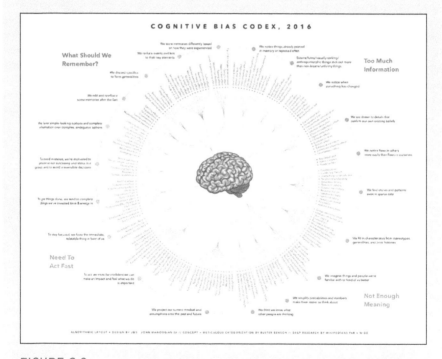

FIGURE 6.9

Buster Benson's Cognitive Bias Codex is a comprehensive, humbling look at dozens of types of bias.

---

2 David Dylan Thomas, *Design for Cognitive Bias* (New York City: A Book Apart, 2020).

3 Buster Benson and John Manoogian III, "Cognitive Bias Codex," https://betterhumans.pub/cognitive-bias-cheat-sheet-55a472476b18

Making It All Work for You

Stakeholder management is tough. It's very hard work to collaborate with your core team of research stakeholders to diplomatically limit the number of people who are involved as you plan and draft your research plan and identify and recruit the right research participants. Identifying research participants is another challenging part of online research; you need to be mindful of bias in the participant selection process. These are essential steps to help the research run smoothly and prevent headaches and rework down the road. Rest assured, content research gets easier from here! The work to fully develop your research questions, run your study, and analyze the results—topics covered in Chapters 7 and 8, "Analyze Content Research Results"—are smooth sailing in comparison.

CONTENT RESEARCH HELPS MARKETING AND CONTENT TEAMS GET ALONG

An Interview with Heather McBride, Senior Content Designer at Expedia

Heather McBride, currently a senior content designer at Expedia, is a former academic who is just a few years into her career in content design. As a relative newbie, she brings a fresh (and refreshing) perspective on the day-to-day challenges in her content work, especially pertaining to stakeholder management.

Heather identifies the friction between content teams and other teams, especially marketing, as an empathy problem. "Marketers are trying to get their own work done—announcements, blogs, press releases—*before* the product or feature is even finished," she says. "Meanwhile, creating good content takes a lot of time! It's therefore not surprising that content and marketing teams can be at odds with each other."

To smooth over the bumps inherent in the marketing-content team relationship, Heather makes sure to involve her marketing colleagues in her team's content work as soon as she can. This includes collaborating on terminology decisions, content research efforts, and any blockers that her team may encounter on the engineering side. That way, the marketing team is as informed as possible about the content—and also has more visibility into and empathy for the speedbumps and challenges that Heather and the content design team are facing.

Heather says the collaboration goes both ways. At her encouragement, the marketing team shares with her all the details and information they have about their many marketing communications deliverables and deadlines. "I share that long list of deliverables with the content team," Heather says. This approach helps the content team understand the stress and pressure that the marketing team is feeling.

This early collaboration and frequent communication results in better alignment of the marketing communications and the product content. There are fewer surprises and terminology hiccups—and, as an added bonus, the teams are also happier and less prone to burnout.

Craft Your Content Research Questions

B y now, you have a solid idea of specifically what you want to learn from your research. You've identified your core stakeholders and worked with them to draft your research plan and audience targeting. Terrific!

You're in the home stretch. Now comes the work of carefully crafting your research questions, to ensure you get the most out of your research efforts. Most of the care you need to take when creating research questions centers on this: doing your best to minimize bias in how your questions are worded, to avoid skewing or overly influencing the answers that your participants provide.

The act of carefully crafting your research questions is quite fascinating, as it requires some knowledge of human behavior. We're all biased and more easily influenced than you'd think! As the research lead, it's your responsibility to format your questions and response options precisely, to make them as "clean" and as free of bias as possible. This gets easier with practice.

In this chapter, we'll review basic question types: multiple-choice, rating-scale questions, and open-ended questions. In Chapter 10, "More Content Research Techniques," you'll learn additional content research approaches, such as 5-second tests, and more involved frameworks like SUS, or System Usability Scale.

Multiple-Choice Questions

Multiple-choice questions are the workhorses of content research. You've probably seen thousands of multiple-choice questions over the course of your lifetime, from taking pop quizzes in school to participating in consumer ressearch. Because multiple-choice questions are familiar to you and your stakeholders, they're easy to lean on and leverage in your content research efforts.

Types of Research That Multiple-Choice Questions Are Good For (with the ideal research types in **bold**):

- **Actionability**
- **Audience-specific**
- Clarity
- Comprehension
- Hedonic or sentiment
- **Naming**
- **Preference**

Output of Multiple-Choice Questions: Quantitative data

Multiple-choice questions tell you *how many* respondents chose which response, or what percentage of respondents chose which response. (See the sample "donut-chart" visual in Figure 7.3.)

Summary of Tips Writing for Multiple-Choice Questions:

As you'll soon see, asking a single multiple-choice question often means a triple-play of sorts. For each multiple-choice question you ask, it's a great idea to include two related follow-up questions, to get the most insights as possible from your participants. Multiple-choice questions can be used for just about every type of research study, although they're especially useful for actionability, audience-specific, naming, and preference studies.

Here's a list of tips to consider as you create multiple-choice questions:

1. Limit the number of response options.
2. Be consistent with the length of response options.
3. Provide a response option for "None of these" or "Other."
4. Ensure that response options are distinct.
5. Avoid "leading" questions.
6. Avoid plain "yes" or "no" response options.
7. Allow participants to skip a question when possible.
8. Immediately follow up with a "tell us why" question.

Limit the Number of Response Options

When you meet with your A team of core content research stake-holders to draft your research study, beware of the "Kitchen Sink" phenomenon, especially as it pertains to multiple-choice questions. You may find that your stakeholders can quickly come up with a dozen or more potential answers to a single multiple-choice question. If this happens, put on the brakes and hold it!

Once again, as lead researcher, it's your job to rein in your team and keep them on track. It's wise to first limit the number of questions. Just because you can ask a ton of questions, it doesn't mean you should. Remember that you need to review all responses and sift through them to find patterns and insights. Don't thwart your own progress with a too-ambitious number of questions in each study.

Next, you should limit the number of potential responses for each multiple-choice question. Aim to keep the number of responses to five or fewer. Limiting the number of potential answers is smart so that you can take action on your research results. Remember, it's OK to run a research study with as few as 5 or 10 participants.

For your multiple-choice question, if you have a laundry list of a dozen or more answers listed for your participants to choose from, you're increasing the likelihood of "noisy" research results, where it's unclear which question, if any, is truly preferred by participants. "Noisy" means it's hard to discern which answer is the most popular and therefore the best one to move forward with to use in your content.

Whittling down a list of 20 potential responses might seem daunting—but take a careful look. As the content expert in your core team of research stakeholders, chances are you'll be able to quickly poke holes in or otherwise completely dismiss some or many of the answers suggested by your stakeholders. For example:

- Maybe a few response options clearly don't align with your company writing style guide. (For example, a word on your brand's "do not use" list is included in a response option.)
- Perhaps the response options don't have the right tone.
- Some responses might include content that's too similar to your competitor's content.
- Some responses may be too similar or duplicative to one another. (Responses should be easily discernible from each other, instead of differing only slightly. If stakeholders are clamoring to include very similar responses, coach them away from that line of thinking.)

Use your broad and deep knowledge of your content standards and principles, your customer needs, and the competitive landscape to whittle down a mile-long list of suggested answers to a manageable, short list.

Be Consistent with the Length of Response Options

As a content researcher, it's your job to make each option in the list of potential answers to a multiple-choice question look equally attractive. To this end, try to make each answer about the same length or character count whenever possible. If you have five potential answers, and four of them are short and about the same length, and then one is extremely long, you're going to see some participants choose the long one simply because it stands out. On the other hand, you're also going to repel a certain percentage of participants who are flashing back to their standardized test-taking days, and they'll avoid the longest response, simply because it's different and jumping out at them. Figure 7.1 shows this phenomenon in action.

Which of the following names do you think is most appealing for Carl's Candy Company's newest chocolate bar?

A. LaLaLaLaLa

B. Supercalifragilisticexpialidocious

C. DoReMiSoFa

D. None of these

(If you selected "None of these," you'll have the opportunity to share any alternate suggestions you have in the next question.)

FIGURE 7.1

A sample multiple-choice question with one response option that's noticeably longer than the others.

Whenever you can, avoid any noticeably long or extremely short "outlier" answers, so as not to inadvertently skew the results. That way, you give each answer a fair, fighting chance to be chosen. When working with many stakeholders from outside your content team, make a point of reviewing each of the response options for multiple-choice questions during collaborative working sessions and edit any obvious outliers as best you can.

Provide a Response Option for "None of These" or "Other"

A best practice for writing multiple-choice research questions is to include a response option of "None of the Above," "None of These," or "Other." This is a fantastic way to get research participants to do some of your content-creation work for you. If, using the candy bar example, you had some participants reply "None of the Above," then by all means, give your participants the chance to express themselves and share what's on their mind by asking them what they would suggest. You may serendipitously find that your research participants come up with some outstanding suggestions that are by far more intriguing than any of those that you or your A Team thought up to include in the initial list of response options.

The easiest way to go about requesting this input from research participants is to provide a follow-up question immediately after each multiple-choice question. If you're using an online research platform, this follow-up question will be in the "open-ended" or "written response" format. Here's an easy way to phrase that question: "Tell us what you'd like to suggest instead. Feel free to provide as much detail as you would like."

Ensure That Response Options Are Distinct

When crafting your multiple-choice response options, make sure that the answers aren't too similar. (See Figure 7.2 for an example of a well-written multiple-choice question.) Otherwise, participants will be confused about which answer to choose.

In Figure 7.2, each of the responses describe "fast," but each response option uses a unique way to describe "fast." Because each response is sufficiently different, research participants ought to be able to discern between them and decide on their response quickly and easily.

Do this:

Which is your preferred name for this new, fast pizza delivery service?

1. PizzaNow

2. Snappy Pizza

3. Insta-Pizza

4. Quick Pizza

5. Other

FIGURE 7.2
In this well-written multiple-choice question, the answers are sufficiently distinct from each other.

In this next example, the multiple-choice responses are just too similar, and to be truthful, they aren't worth your time and effort to test. This is an extreme example, but as you ramp up your content research program and work with stakeholders, you'll find that it's easy to fall into the trap of having stakeholders influence a study's research questions too heavily. You need to incorporate their input, but retain editorial control and not let stakeholders dictate the exact wording of the questions themselves.

Which video title on the topic of cooking, if any, do you find most compelling and interesting?

Don't do this:

1. Chicken Cooked Many Ways
2. The Many Ways to Cook Chicken
3. Several Ways You Can Cook Chicken
4. Other (please provide your own suggestion)

If, instead, the responses were substantially different, then you'd have a worthy test on your hands. The responses to this question will tell you which cooking technique is most compelling to your audience, and if you're running a cooking website or app, that would be information worth having.

Which video title on the topic of cooking, if any, do you find most compelling and interesting?

Instead do this:

1. Chicken **Barbecued** Many Ways
2. The Many Ways to **Fry** Chicken
3. Several Ways You Can **Sautée** Chicken
4. Other (please provide your own suggestion)

Avoid "Leading" Questions

Notice how in the question in Figure 7.2, the word "fast" appears in the question, but it is *nowhere to be found* in the response options. That's a best practice! If "fast" were to appear both in the question and in one (or more) of the response options, that would be an example of a *leading question*. And yes, you guessed it, a leading question can skew or unfairly influence the responses and render your research less effective than it could be.

Leading questions can crop up when you write your test plan too quickly, or if you decide to work as a lone wolf and skip the important step of asking stakeholders to review your test plan. (Which, as mentioned in Chapter 6, "Research Planning and Stakeholder Management," is not recommended.)

There are other ways to inadvertently create leading questions. For example, if you ask participants, "Which of the following options do you like best?" then you're *forcing* them into liking one of the options. What if they don't like *any* of them? That's quite possible. You need to give participants an out. Be careful to word your questions in a *neutral* way, to avoid bias as much as possible.

Avoid Plain "Yes" or "No" Response Options

If you're mindful to avoid plain "yes" or "no" response options for multiple-choice questions, then participants won't perceive that there's a "right" or "wrong" answer. This is a tactic that's always a good rule of thumb for interviewing anyone, anywhere. (It's also an approach that will be familiar to all of you former journalists out there who now work as content professionals.) It's preferable to frame your multiple-choice questions so that participants can choose from specific, unique response options. Including an "Undecided" or "Maybe, I'm not sure" response option is always a good idea.

Here's an example of a well-written multiple-choice question that avoids including only "yes" or "no":

Do you think you would be likely or unlikely to order a pizza from this restaurant?

1. Likely
2. Unlikely
3. I'm not sure

Allow Participants to Skip a Question When Possible

If possible, allow participants to skip questions. (Note that not all online testing or survey platforms are flexible about this.) Sometimes, despite your best efforts to screen participants and to provide enough context and clarity for the questions, participants just won't feel comfortable or confident answering a question. If you're able to allow participants to simply skip a question when needed, doing so is helpful for boosting the reliability of the responses. If you find a high percentage of participants are skipping a specific question, then you probably need to go back to the drawing board and work on improving the clarity of that question.

Immediately Follow Up with a "Tell Us Why" Question

Another best practice is to include an open-ended follow-up question immediately after each multiple-choice question. In this question, ask the participant:

> Please explain your response to the preceding question.

> or

> For Question #X, please tell us a bit about why you selected the response that you did. Walk us through your thinking, providing as much detail as you would like.

This open-ended or written-response question format, where the participant gets to speak or enter their response and provide as much or as little detail as they want, can be used to collect qualitative information that can be very enlightening. The one limiting factor with this question format may be your online research platform, if it has a word-count limit for these open-ended questions. If that's the case, explicitly call out in your question how many words or characters that participants are allowed—including spaces! Some online platforms will provide a word- or character-count countdown that displays the remaining available characters.

The gentle encouragement you provide in your "tell us why" question will help people who have a hard time expressing themselves to relax and loosen up a bit, given that they now know that you really, truly want to know what they're thinking and that you value the information they have to share. Notice how friendly that sentence is: "Provide as much detail as you would like." It's freeing, encouraging, and much more easygoing compared to "Make sure to give lots of details!" or something similar. You'll be more successful in generating qualitative feedback with gentler phrasing.

Sample Multiple-Choice Question Output

Multiple-choice questions provide quantitative insights and are especially easy to interpret when you use online research tools like UserTesting, dscout, or UserZoom. When you ask a multiple-choice question, the responses are tallied, and the output is a colorful chart (formally called a *donut graph*) that clearly shows the breakdown of responses (see Figure 7.3).

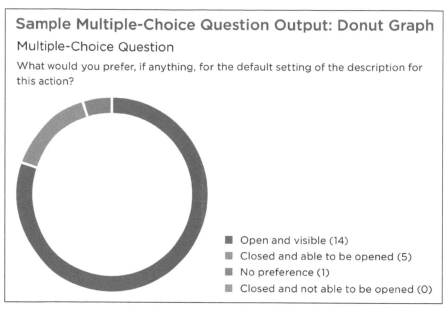

Sample Multiple-Choice Question Output: Donut Graph

Multiple-Choice Question

What would you prefer, if anything, for the default setting of the description for this action?

- Open and visible (14)
- Closed and able to be opened (5)
- No preference (1)
- Closed and not able to be opened (0)

FIGURE 7.3

A sample output from a multiple-choice question (donut graph) for a usability study. In this instance, the most popular response was chosen nearly three times as often as the next most popular answer.

If, following your multiple-choice question, you opt to include a "please tell us why" follow-up question, that will provide you with valuable qualitative insights. See Chapter 8, "Analyze Content Research Results," for step-by-step guidance on how to pan for gold and uncover insights in qualitative research question responses.

Rating-Scale Questions

Rating scales are also known as *Likert scales,* after the psychologist who is credited with inventing them, Rensis Likert. Like the name implies, they help gauge the responses of your research participants along a scale or range. Here's an example:

On a scale of 1 to 5, where 1 represents Not at All Likely, 3 represents a Neutral response or no opinion, and 5 represents Extremely Likely, how likely or not do participants feel they are to take a specific action?

Scale or Likert questions are immensely versatile in content research, as they can be written with an endless variety of different "endpoints."

Most often, scale questions are written using an odd-numbered range, with a maximum of 9:

> On a scale of 1 to 9, with 1 being Not At All Likely and 9 being Extremely Likely, how likely are you to recommend our app to others?

This limited scale range helps research participants make a decision more easily. (Can you imagine how much mental gymnastics you'd have to do to respond to a question with a scale of 1 to 50 or 1 to 99?)

Types of Research That Rating-Scale Questions Are Good For (with the ideal research types in **bold**):

- **Actionability**
- Audience-specific
- Clarity
- Comprehension
- Hedonic or sentiment
- Naming
- **Preference**

Output of Rating-Scale Questions: Quantitative insights

When using an online platform, you'll get easy-to-read charts or graphs, such as bar charts, also known in the research world as *histograms*. These are simple types of data visualizations that make it easy to interpret and understand the research results, which in turn help you easily take action to update and improve your content.

Summary of Tips for Writing Rating-Scale Questions:

1. Use an odd number of points or steps in the scale.
2. Use low numbers on the scale for negative sentiment and high numbers for positive sentiment.
3. Be consistent with endpoint descriptors.
4. Include both negative and positive endpoint descriptors in your questions.
5. Avoid "leading" questions.

Use an Odd Number of Points in Your Scale

Most online research platforms provide a variety of sizes of scales, usually 5, 7, 9. Using an odd-numbered scale sets you up for success. It provides a true midpoint, so you can easily determine which way the wind is blowing: that is, are most responses skewing toward one end of the scale or the other? Of course, it's possible, although the probability is low, that you may some day see a scale question where responses fall smack-dab in the middle of the scale, which represents that participants are neutral and don't really have an opinion one way or another.

Use Low Numbers for Negative and High Ones for Positive Sentiment

In a Likert or scale question, the number 1 should always be the "pain" point of the scale, representing dislike, difficulty, and so on. The midpoint is neutral. And the highest number, whether 5, 7, 9, or some other odd number, is the most positive point on the scale. This is a standard best practice.

If you include a series of scale questions within a single research study, don't confuse participants by writing one question with "1" as the negative side of the scale and a later question with "1" as the positive side of the scale. That's an easy way to introduce cognitive load into the research process and therefore noise into your research results.

Be Consistent with Endpoint Descriptors

Always be consistent with your phraseology. For example, if you want to know how clear or not your participants think a certain word or phrase is, use "Not At All Clear" for the most negative response and "Extremely Clear" for the most positive response.

Note that the most negative response should *not* be written as "Completely Unclear." That's because "completely" is a positive word. Mixing the positive word "completely" with the negative word "unclear" isn't quite an oxymoron, but it's definitely a confusingly worded phrase. In the user experience field, "completely unclear" is a phrase that can be described as "high friction" or "high cognitive load," because it requires some mental gymnastics to understand. It's what is called *friction-filled phrasing*.

These recommendations about wading carefully as you write your research questions may feel nitpicky at first. But think about it: If you have a research study that includes just five or so questions, you want to do all you can to make each question crystal clear. You need to make sure that your participants can read and understand each question easily and respond accurately. You also want to hold the attention span of your participants, so that they're giving as much energy and attention to your last question as they do to the first one.

Include Both Negative and Positive Endpoints in Scale Questions

As you write scale questions, be sure to include labels for both the negative and positive endpoints of your scale, being carefully consistent with your descriptors as you go.

To avoid introducing bias into the response, it's also best to mention both ends of the scale in the text of each research question.

For example, here's an Actionability research question in scale format:

> Given the screenshot of the app you just saw, how unlikely or how likely do you feel you are to download this app in the near future?

Including both endpoints of the scale within your question reinforces them in the minds of your participants. It also lowers the chance that a participant may misread your question and reply in a way that doesn't accurately reflect how they feel.

PRO TIP CAPITALIZE AND BOLD THE ENDPOINTS

Remember that capping and bolding the endpoints can be helpful for participants, to assist them in quickly comprehending the question at hand.

Avoid "Leading" Questions

Just as with multiple-choice questions, it's easy to inadvertently inject bias into scale questions. For example, for a hedonic or sentiment question, you wouldn't want to ask:

> On a scale of 1 to 5, where 1 represents Good and 5 represents Excellent, how good do you feel when you hear these song lyrics?

In that question, you're asking for only positive responses.

Similarly, if you're writing a comprehension question, you don't want to ask:

> On a scale of 1 to 9, how easy is it for you to understand the words in this app?

If you write a research question in this sort of leading way, you're cornering your participants and not giving them the chance to respond honestly. Unfortunately, you're also going to skew your research results and render any "insights" from such questions less accurate, reliable, and actionable. To reduce the bias in this question, you need to include both the positive and negative endpoints:

> On a scale of 1 to 9, where 1 is Very Hard to Understand and 9 is Very Easy to Understand, how would you describe the words in this app?

An additional note here about scale questions: keep them simple. Don't try to squeeze two topics or questions into one. It takes time for participants to absorb all the information, endpoints, and details in a scale question. Don't make their work harder by combining too much information into a single scale research question. By keeping them simple, you'll also make your own work easier, when it comes to analyzing results.

Sample Scale Question Output

Figure 7.4 shows a sample output from a rating-scale or Likert research question.

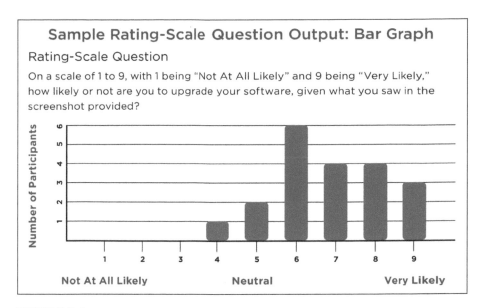

Sample Rating-Scale Question Output: Bar Graph

Rating-Scale Question

On a scale of 1 to 9, with 1 being "Not At All Likely" and 9 being "Very Likely," how likely or not are you to upgrade your software, given what you saw in the screenshot provided?

FIGURE 7.4

This sample output of a rating-scale question (bar chart or histogram) shows that respondents are mostly moderately or very likely to take the action in question (upgrading their product).

Open-Ended Research Questions

Qualitative, open-ended question formats are extremely useful for uncovering golden-nugget insights about your content. Simply put, open-ended questions are those where you let participants tell you what's on their mind. As you've read earlier in this chapter, it's a smart one-two punch to include an open-ended question *immediately* following each multiple-choice question. Open-ended questions are also powerful on their own and serve as a safety net of sorts, allowing you to make sure that you haven't missed any major topics or points that your participants may be thinking about. ("Anything you'd like to add?" is a great question to include at the end of each research study.)

Responses to open-ended questions do take much more time to analyze, compared to quantitative research approaches like multiple-choice and scale questions. But that time is well spent—an investment in improving your content and customer experience.

Types of Research That Open-Ended Questions Are Good For (with the ideal research types in **bold**):

- Actionability
- Audience-specific
- Clarity
- **Comprehension** (open-ended questions are probably the best format for this type of research!)
- **Gap and excess analysis**
- **Hedonic and sentiment**
- **Preference**

Output of Open-Ended Questions: Qualitative data, or information and details about *why* your research participants are thinking the way they are.

Also, direct quotes, sometimes called *verbatims*, from participants, which can be very powerful in case studies and slide presentations about content reseach and results. For examples of qualitative research outputs, see Chapter 8, "Analyze Content Research Results," including the section, "Analyzing Open-Ended (Qualitative) Responses."

Summary of Tips for Writing Open-Ended Questions:

1. Use as neutral phrasing as possible.
2. Don't start with an open-ended question.
3. Encourage people to provide as much detail as they'd like.
4. Include an "Anything else?" question at the end of each study.

Use as Neutral Phrasing as Possible

Just like you saw in multiple-choice and scale or Likert questions, you need to be careful to avoid "leading" your respondents. You may have a hypothesis or hunch about your content. You may be running a naming study and feel absolutely convinced that you know the "best" name for a product or feature, and you *just know* the study will prove you right. Or you may be conducting a comprehension study and feeling super confident that your participants will understand your content as it's currently written. Or you might be in a big hurry and simply become a bit sloppy when writing your research questions.

Whatever the scenario, do your best to write your questions neutrally, to avoid "giving away your cards." Remember that research participants, whether you're working with them online or in-person, tend to want to please you. You and your business are depending on your research to help inform potential changes to your content and customer experience. Before starting your research, pause and read your open-ended questions aloud, to help minimize any biases.

Don't Start with an Open-Ended Question

Starting off your study with an open-ended question isn't typically a good idea. Your participants haven't "warmed up" yet, so they are unlikely to feel confident about sharing their thoughts. Start with a few "softballs"—a multiple-choice question, a scale question, or maybe a task-oriented question where you share a screenshot or prototype of your content and ask specific questions about it. That way, when you ask an open-ended question that requires either some writing or talking, participants will feel more comfortable and more likely to speak candidly about what's on their mind—providing you with valuable insights and helpful information.

Encourage People to Provide Details

There are many reasons why you need to be as encouraging as possible when it comes to qualitative questions. As mentioned previously, people can be reticent about replying to open-ended questions. If they haven't participated in research before, they may not know what to expect and feel nervous. You will also probably have some introverted people participating in your research. In addition, people who identify as neurodiverse can benefit from encouragement during the research process. To get people to talk (or write) when responding to open-ended questions, you need to provide them with a gigantic green light and warm encouragement.

This is an especially important point to absorb for all you content pros. Writing is your jam. For many people, writing is anxiety-triggering and something they tend to avoid!

You need to make participants comfortable and set the stage for them to open up. On every open-ended research question, it's helpful to tack on this sentence: *Please provide as much information or detail as you would like.*

That way, you'll be more likely to get full-sentence responses, instead of an abrupt word or two. And don't be afraid to repeat that encouraging sentence for every open-ended question. It may feel redundant and unnecessary, but it makes a huge difference.

Include an "Anything Else?" Question

For your very last question in each research study, consider adding an "Anything else?" question. "What else would you like to add?" or "What topic or detail that wasn't mentioned in the previous questions would you like to bring up and comment on?" (Note that these are not phrased as questions with mere "Yes" or "No" answers; they're deliberately phrased to help coax your participants into opening up and sharing.)

This last catch-all "safety net" question often elicits answers that lead immediately to additional content research! Remember, you are not your customer or audience. You don't know what you don't know, and this last "What else?" open-ended question will help you better understand your customers and what they're thinking about when it comes to your content.

> **NOTE** ACCESSIBILITY AND OPEN-ENDED QUESTIONS
>
> Some people are less inclined to type out their responses or have accessibility issues that make this difficult. When using online research or survey platforms, try to aim for a balance of written and verbal response questions. As mentioned previously, a lot of people simply don't like to write. Others are self-conscious about grammar or spelling, and therefore may tend to keep their written answers on the short side.
>
> Verbal response answers often result in even more "golden nugget insights" than written responses because people are thinking out loud and immediately verbalizing the thoughts that quickly pop into their mind. Fortunately, verbal answers have become easier to analyze, as many online platforms "translate" them into transcriptions.

Gauging Participants' Reactions

Open-ended questions are also helpful when you're able to view your participants' facial expressions and body language as they read your questions aloud. (Some online research platforms also record videos of participant sessions for you to watch and analyze.) Do your participants appear amused, even entertained, or confused and lost, or pained and emotional? It can be very helpful to share video clips or screenshots of research participants' reactions with your colleagues and research stakeholders, to reflect just how unclear—or how wonderfully effective—the content is.

That said, there are ethics involved here. Some industries—health-care or finance, for example—can drum up painful experiences and even traumatic memories. Proceed with care and be respectful of participants' past histories and experiences. Let them know they don't need to complete the research questions and can bow out at any time, should they feel uncomfortable or don't want to continue for any reason.

Making It All Work for You

Multiple-choice questions, scale (or Likert) questions, and open-ended questions are the content research workhorses. If you learn how to craft only these three types of questions, you'll be able to uncover immensely helpful insights and "golden nuggets" for you and your content team. We'll get into how to analyze their results in Chapter 8, "Analyze Content Research Results." There are, of course, dozens of potential question types—more of those will be covered in Chapter 10, "More Content Research Techniques."

LEVERAGING CONTENT RESEARCH TO SUPPORT PLAIN LANGUAGE

PHOTO CREDIT: KARINA MORA

An Interview with Michael Metts,
Principal Content Designer at Expedia

As a principal content designer at Expedia, Michael Metts is also a frequent speaker at content-focused conferences like Button and the coauthor with Andy Welfle of *Writing Is Designing*, another title from Rosenfeld Media.

He answered some questions about working in content design, getting started with content research, and the importance of plain language.

Q: What has been your experience with content research?

A: [In a previous role], we did highlighter testing of emails—Qualtrics now has highlighter testing—with a large sample size. We had 700 people, 350 for each variant. We were curious about the use of plain language compared to jargon and words that were longer, with more syllables. We had one half of the people highlight where the copy was helpful and the other half highlight where it was unhelpful—as in, unclear, or they felt it was jargon.

continues

We found that a lot of people preferred simpler language, like "Talk to someone" versus "Contact an associate." It helped make the case to roll out plain language across the organization, which was pretty great. It gave us a rationale for our [content] standards. This is why we do [content design], right? My team and I were like, "Let's go!" We knew it would have an impact.

Q: I hear often that people (erroneously) equate simple or plain language with "dumbing down." How do you react to comments like that from stakeholders that don't (yet) understand what plain language is?

A: Sarah Winters has been talking about it for a long time—plain language.

I worked for an insurance company, and it was B2B [business-to-business]. We were selling to businesspeople. Stakeholders therefore thought we had to use "business language" when talking to these customers. You can see how they might think that. And there were some customers who understood all the jargon, you know, people who'd been customers for 15 plus years—we called them "super users." And they had learned all the terms like "adjudication."

But simple language just helps you understand things faster. We can [be clear], and we should. It has a lot of accessibility and usability benefits.

Q: Do you have any advice for people who are starting out with content research?

A: Yeah, keep in mind you can plug content questions into usability tests [being run by your UX research team]—and tack them on to the beginning of a usability testing session. You can also do it yourself pretty easily. You can ask your audience questions like, "What do you think [this word] means?" and "What do you think will happen next"? There's so much opportunity to learn what people are perceiving, reading, and understanding.

CHAPTER 8

Analyze Content Research Results

B y carefully creating your research studies, you're setting yourself up for success and should be well prepared to analyze the results as they're available. This is the most exciting part of research— time to uncover the golden nuggets! Remember, research that's quantitative, or answers the "what" questions, is quicker and easier to analyze. The qualitative, or "why" research, takes more time and effort to wade through—and it's often the qualitative research that's likely to reveal astonishing, surprising information about your audience.

Whether you use an online platform or conduct research in person or using email or other methods, it's relatively easy to analyze your quantitative research questions. That said, there are a few potential "gotchas" to be on the lookout for, to help ensure that you take the best next actions based on participant responses:

- Remember to keep in mind that the results you see from your research responses are *not* statistically significant. *It's fine that they're not!* You may hear grumbling from stakeholders who want statistically significant data. If they really want to conduct an A/B experiment informed by your research and wait days or weeks for it to reach statistical significance, so be it. But you can gently remind them that the information you gleaned from your content research is actionable by itself, without the added layer of time and effort.

- Please avoid referring to the most frequent responses as if they're the be-all and end-all, set-in-stone "facts" about your content or customers. Your research is a snapshot of information gathered at a specific point in time, not written in stone. Your results are actionable, although they should be revisited periodically in the future to ensure that they reflect customers' preferences or are updated as needed.

- Most importantly, please avoid using the word "winner" for the most frequently seen responses in your research. This word demeans the other contenders, which you and your stakeholders worked hard to identify.

Analyzing Multiple-Choice Questions

Multiple-choice questions are simple to analyze, especially if you were careful to take the time to ensure that each response option is sufficiently different from the others. Multiple-choice questions provide quantitative data, that is, numbers (how many people out of the total number of participants responded a specific way to the question).

Unambiguous Multiple-Choice Results

Say you're curious what to name a new feature. You ran a study with 10 people from your target audience as participants. The responses can be summarized like the examples listed in Table 8.1. In this case, your next steps are obvious, since Response 4 was chosen by a clear majority of participants. Go forth and use the content from Response 4 in your customer-facing content (but do so diplomatically and carefully, by referring to the advice in Chapter 11, "Apply Insights and Share Business Results").

TABLE 8.1 EXAMPLE OF RESEARCH RESULTS WITH CLEAR-CUT NEXT STEPS

Response Number	Number of Study Participants Who Chose This Response
Response 1	1
Response 2	2
Response 3	0
Response 4	6
Response 5	1

The results in Table 8.1 represent what can be called a *slam dunk*. Based on the responses from the target audience, it's clear that Response 4 appears to be the most popular with 6 responses out of 10. Three times as many people chose Response 4 as the next most popular answer, which was Response 2. You should feel confident about moving forward and using the feature name that appears in Response 4. (And obviously, you should avoid implementing the content in Responses 1, 2, 3, or 5.)

A quick side note: It's notable that Response 3 has no responses at all. *What's going on here?* Your Spidey Senses should be tingling. If you have time to dig further, it may be worth a second follow-up study to find out what exactly is so unattractive about this specific response. Knowing *what* content is *unappealing* to your audience and *knowing why* can be just as powerful as knowing what content *is appealing and engaging* to your audience and why.

What about when research results aren't quite so clear-cut? Table 8.2 provides an example of research responses that are slightly unclear and require a bit of patience and sleuthing to interpret and decide on next steps (that is, determining which response, if any, ought to be used to inform or update your customer-facing content).

TABLE 8.2   RESEARCH RESPONSES *WITHOUT* CLEAR NEXT STEPS

Response Number	Number of Study Participants Who Chose This Response
Response 1	1
Response 2	3
Response 3	0
Response 4	5
Response 5	1

The results in Table 8.2 are not a slam dunk, unlike Table 8.1. Clearly, however, it's best to steer clear of Responses 1, 3, or 5, as they appealed to no or few participants. It's also clear that Response 4, with its five responses, is the one whose content should be strongly considered for implementation or publishing.

Response 2, with three responses, is the next most popular response. If you were to take Response 2 and implement that in your customer-facing content, this wouldn't be the end of the world. It may not be quite as engaging as Response 4, but it could probably still pass muster.

In a situation like this, with two responses bubbling to the top, you could do a "runoff" if you're so inclined, and you and your team have the time to do so. What this means is you'd run an additional research study, using the same questions as in the original study, but using only two potential responses: Response 2 and Response 5. (They would then become Response 1 and Response 2.)

However, you could probably (hopefully) avoid the need for a runoff, if you followed the formula for the "one-two punch" of writing a multiple-choice question and then immediately following it with an open-ended question that asked, "Tell us some details please about why you chose the response you did." If you included the open-ended follow-up question, you'd have the related qualitative responses to analyze as well. (See "Analyzing Open-Ended (Qualitative) Responses" later in this chapter on how to review and distill open-ended responses and uncover insights from qualitative research questions.)

When you have less than clear-cut responses to a multiple-choice question, once you carefully evaluate the responses to the related open-ended qualitative ("why") questions, a runoff is often unnecessary. That's because the answers to the open-ended questions are often chock-full of fascinating information and details about your participants' thoughts and opinions that will stop you in your tracks and prompt you to say, "Aha!" By understanding the rationales behind *why* your research participants selected the choices they did, you'll invariably have some golden-nugget insights to help with your decision-making. You'll be able to determine which research response is the one to "run with" to make your customer-facing content as strong as it can be.

Ambiguous Multiple-Choice Results

There will be times when you get a "painful split"—a sprinkling of responses among the options, such that there's no clearly outstanding response to run with. This happens! It may be because you and your stakeholders came up with brilliant response choices that appealed equally to your research participants. Or it may be because the responses weren't sufficiently different from each other. Perhaps the root cause could be that the question or responses weren't worded clearly enough. (This is why doing a "dry run" or mini pilot test before running your study with actual participants can be a smart move.)

When ambiguous responses occur, there's no need to despair. This is where your experience, knowledge, and discerning judgment as a content expert come in. You can do one of three things:

- **Option A:** Make an executive decision and choose one of the top responses and implement it in your customer-facing content. (Remember your RACI chart: You are the person ultimately responsible for the content decisions. You can make this call on your own, or loop in your core stakeholders to weigh their recommendations. If you do the latter, prepare for some debate around the decision making.)

- **Option B:** Go back to the drawing board and rewrite the question, the responses, or both, and run a fresh test.

- **Option C:** If it makes sense for your team, and you have the resources to do so, run an A/B experiment with the two most popular responses (or a multivariate experiment that includes the two most popular responses). (See Chapter 5, "Identify Your Content Research Goals.") By "if it makes sense for your team," that means four things:

1. **Content traffic or visits:** The content in question—web page, app, or whatever format of content you're evaluating—receives a sufficient daily or weekly volume of customer visits or "hits." In other words, the experiment will be able to reach true statistical significance without you and your stakeholders needing to wait weeks or even months for this to happen.

2. **Patience:** Your stakeholders have the patience to wait for however long it takes for the experiment to reach statistical significance. This could be weeks, sometimes months, for content with limited traffic. As they sometimes say in the digital product world, "the juice may not be worth the squeeze," or the effort may not be worth it.

3. **A/B experiment management:** Your content team, production team, or software development team have the time and know-how to set up, test, run, babysit, and analyze the A/B experiment, and communicate the results effectively (if your content management system is unable to support A/B experiments). If you're fortunate to have a CMS that enables A/B experiments, you as the content lead have the time yourself to do this work.

4. **Gatekeeping and boundaries:** While the experiment is running, you as the content lead need to be prepared to be peppered by stakeholders who will be frequently asking you, "Is the experiment cooked yet? Where are the A/B experiment results? Can you share them yet?" Remember, it's dangerous to share any content research or experiment results prematurely.

If you're thinking, "Sheesh, that's a...lot! Maybe I should just go ahead and choose one of the top responses from the original research study and skip the idea of A/B experimentation," then, yes, you're onto something! A/B and multivariate experiments can be powerful, but they're also absolutely an investment in time, patience, and team-dynamics management. It's often much, much easier to choose one of the top responses from an ambiguous research study over an A/B or multivariate experiment.

Data Visualizations for Multiple-Choice Questions

If you use an online research platform like UserZoom, dscout, or UserTesting, once your study is complete, the responses to your multiple-choice questions will be beautifully and conveniently summarized in a colorful pie chart or "donut graph" that's easy to share. These colorful data visualizations tend to get shared and forwarded a lot in email and chat messages because they so clearly and quickly tell a story about customer engagement.

Figure 8.1 is another example of a donut graph. This particular donut graph is from a multiple-choice question that sought to determine whether participants preferred to be shown some instructional details within the user interface to always be visible, or if they preferred that the instructions be revealed upon selecting a tooltip.

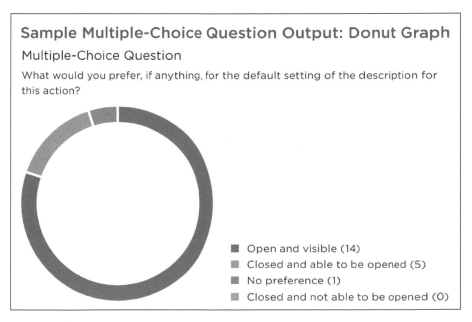

Sample Multiple-Choice Question Output: Donut Graph

Multiple-Choice Question

What would you prefer, if anything, for the default setting of the description for this action?

- Open and visible (14)
- Closed and able to be opened (5)
- No preference (1)
- Closed and not able to be opened (0)

FIGURE 8.1
In this donut graph, which is an output from a multiple-choice research question, there's a landslide response. The most preferred response (with 14 out of 20 respondents selecting it) was nearly three times more popular than the second most chosen option.

If you don't have the benefit of an online research platform and you're manually tracking your results—that is, organizing the responses and tracking them in a document or spreadsheet—you can easily create your own simple chart like the one in Figure 8.1. Or you can build a pie chart from Microsoft Excel, Google Sheets, or a similar app.

There are many online tutorials and YouTube videos with instructions for how to build a pie chart, but basically it consists of these steps:

1. Start with a two-column table, where the first column is the label, and the right-side column is the value.

2. For Google Sheets, highlight your two columns, go to the Insert tab, and then select Chart.

3. For Excel, highlight your two columns, go to the Insert tab, select Chart, and then select Pie.

NOTE ABOUT ACCESSIBILITY AND RESEARCH RESULTS

Make sure that your research results, especially data visualizations, are clearly labeled with clear units of measurement. That way, your research results won't be misconstrued or misinterpreted.

A quick note about accessibility: Globally, an estimated 7 to 8 percent of adults are red-green color-blind, and an additional 1 to 2 percent have other forms of color blindness.[1] Keep this in mind when sharing data visualizations from your research efforts. (Don't use only color to communicate—make sure that your charts have labels and legends so that the information is clear.)

In addition, whenever you're sharing data visualizations in a document or presentation, it's also essential to add alternative text, often referred to as *alt text*, to your slides that clearly but briefly summarizes what's shown in your chart or graph. (Alternative text is the text that appears on an app website to help people with low vision or blindness understand—sometimes with the use of screen reader tools—what visuals are being displayed.) For more information on alternative text, check your apps: Google Docs, Google Slides, Microsoft Word, and Microsoft PowerPoint all offer step-by-step guidance for creating alt text. For more help on alt text, Deque, an organization known for in-depth training courses and guidelines on accessibility best practices, has a helpful guide called "How to Design Great Alt Text."[2]

---

1 Colour Blind Awareness, "About Colour Blindness," https://www.colourblindawareness.org/colour-blindness/

2 Deque, "How to Design Great Alt Text," https://www.deque.com/blog/great-alt-text-introduction/

Analyzing Rating-Scale Questions

Like multiple-choice questions, scale questions are straightforward to interpret and summarize (see Figure 8.2). Analyzing scale questions is similar to the previous process outlined for multiple-choice questions, as the output for both question types is quantitative data—the basic information or "what" of your research findings, such as what content people prefer, what content makes them feel a certain way, or what actions they would take after reading your content, and so on.

The output for a scale question is typically a *bar chart*, also known as a *histogram*. Please, please, refer to them as plain old bar charts. Remember the recommendations in Chapter 2, "Leverage the Power of Clear Content," about the wonders of plain language and the pain and confusion induced by jargon and unfamiliar words. Simple language is powerful language because it's easy to understand.

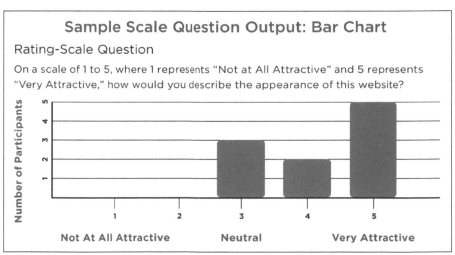

Sample Scale Question Output: Bar Chart

Rating-Scale Question

On a scale of 1 to 5, where 1 represents "Not at All Attractive" and 5 represents "Very Attractive," how would you describe the appearance of this website?

FIGURE 8.2

A sample output from a rating-scale question, showing a majority of responses are positive. Three participants were neutral, two agreed that the website was attractive, and five respondents strongly agreed.

When analyzing scale question results, you should be in good shape if you used a scale with an odd number of points (scale of 1 to 5, 7, or 9), and if you wrote your questions using clear, straightforward endpoints.

With an odd number of points along your scale, you'll have a clear midpoint, and you can easily note which way the responses lean. Are most responses toward the negative or the positive end of your scale? This is valuable information in and of itself. Then evaluate which specific number or point along the scale has the most responses.

Just like multiple-choice questions, sometimes you'll find that it's crystal clear which response stands far out from the rest, and that it obviously makes sense to use its content to be incorporated into your customer experience. Sometimes, however, you'll find there's a response that's chosen by the greatest number of participants, but that the total number of responses is not all that much greater than the next most popular answer.

Remember again that the goal of this type of quantitative content research is *not* to reach statistical significance, but to simply carve a path forward to help you and your team ensure that your content is reflecting feedback from your customers. (Remember, you're working on website content, not developing new cures for cancer.) Don't break into a sweat if you have quantitative research responses that are neck-and-neck with each other. Just like with multiple-choice questions, it's usually perfectly fine to pinpoint the response chosen by the greatest number of participants as the one whose content you should implement. And, if the stakes are quite high and you see that two or even three responses were selected by nearly the same number of participants, you can always do a runoff, if time allows.

Analyzing Open-Ended (Qualitative) Responses

Analyzing open-ended or written-response questions and their qualitative responses takes significantly more time than evaluating quantitative research answers. Keep in mind that you get out of it what you put into it. While multiple-choice and scale question responses typically result in data that you can quickly analyze within just a few seconds or minutes, distilling qualitative answers into insights requires an investment of patience, perseverance, and time to comb through them.

Reducing Bias in Qualitative Evaluations

You may want to divvy up the work when it comes to analyzing open-ended questions so that many hands make light work. Depending on how many participants were involved in your study, there can be lots of lengthy responses to wade through, and, depending on your testing platform, you may have many videos to watch as well. Bringing in other stakeholders to evaluate responses is also a hedge to prevent the biases that are inherent in every human being's mind, so that they don't get in the way of interpreting the results. That said, be thoughtful about whom you choose to help you out.

You probably have several stakeholders on your A Team who were involved in the planning and setup of your research study. It's tempting to tap them to help analyze the open-ended question responses. After all, they're familiar with your research plan, hypothesis, and goals. However, pause for a moment! When it comes to analyzing qualitative question responses, it's all too easy for one's bias to infiltrate the analysis process. If one of your stakeholders has an agenda for the research—articulated or not—and is hoping to see a specific result from the research, their analysis will be biased, as they are consciously or unconsciously hunting for information that supports their predicted or desired research outcome. Qualitative analysis demands as much neutrality as possible, to accurately understand the thoughts that customers are expressing and not to cherry-pick the feedback that you're expecting or hoping to hear.

For these reasons, if you're able to recruit enough colleagues who have been trained in or are familiar with research best practices *but who were not involved at all in the initial research planning*, it's probably a wise choice to have them be your helpers to distill the qualitative feedback to uncover insights and actionable information.

Reviewing and Categorizing Qualitative Feedback

Some people prefer to analyze all open-ended responses from a given research study in one fell swoop. However, if you're using UserZoom, dscout, UserTesting, or a similar platform, you can start analyzing data as soon as each participant's response is received. This latter approach can be helpful in reducing bias, because the person or people doing the analysis won't be as easily influenced by the information and details seen within the other participants' responses. Similar to planning and writing research studies, distilling and analyzing research study results is a muscle that grows the more you use it. As you gain more experience, you and your team will gain a sense for what works best for you. If time is of the essence and you need to distill your research as quickly as possible, check out the tips mentioned in the interview in this chapter with experience design researcher Irish Malig.

Here are four basic steps for evaluating qualitative research responses:

Step 1: The First Read

The first step in searching for insights and golden nuggets in qualitative responses is to simply read the responses slowly and with care. (In the case of video or audio recordings of research, watch or listen mindfully.) Just take a first stab at reviewing the responses, to get a general feel for the information in front of you. Do this first review in whatever format works best for you and how you absorb information. This may involve reviewing completely online within your online research platform, exporting responses to Excel, or printing out the responses. You do you.

Step 2: Highlight the Most Notable Comments

The next step is to re-read (or listen to, or watch) the responses and highlight the comments and information that jumps out at you. Highlight notable comments that grab your attention and make you say, "Whoa," or "Wow!" or "Hmm, that never occurred to me." You can start to distill the responses with the help of color-coded highlighting or various colors of sticky notes.

You may find that you immediately notice several participants responded to a question in a similar way. Patterns or themes like this are good to mentally note in this step, but don't get preoccupied with them just yet; that's what Step 3 is for!

Step 3: Categorizing or Affinity Mapping the Responses

Here's the fun part: Categorizing the highlighted comments or sticky notes. This is also referred to as *tagging* or *coding* the information in the responses. It's also known as *affinity mapping*, or the *KJ Method*, named after Jiro Kawakita, an ethnographic researcher credited with developing the technique in the 1960s.

It's up to you what system you want to use for this process, although it differs slightly whether you are using your laptop or if you go "offline" to use printouts. If you're marking up a Microsoft Excel or Google Sheets document, you may choose to use a unique color of highlighting or font for each participant and add an additional column in the spreadsheet for noting what you determine to be the overarching category or classification of each highlighted word or phrase.

(If you're dividing the labor of qualitative analysis among yourself and several colleagues, make sure to coordinate among yourselves about who is using which color, and for which purpose!)

Say you're creating a music app. You run a content study conducted using an online research platform with 20 participants. The first question is a multiple-choice question, asking participants which singer is the best of all time: Aretha Franklin, David Bowie, Dolly Parton, Prince, or Someone Else/None of the Above.

You smartly followed up this question with the handy, best-practice, open-ended questions of:

If you replied, "Someone else," please share who that is.

And then:

Tell us a bit about why you chose the response you did. Provide as much detail as you'd like.

The multiple-choice breakdown of results were as follows:

- Aretha Franklin: 4
- David Bowie: 3
- Dolly Parton: 4
- Prince: 5
- Someone else: 4

For the "Someone else" responses, the musicians mentioned were Adele, Christina Aguilera, Stevie Nicks, and John Lennon.

So, in this case, the multiple-choice responses were pretty much neck-and-neck, without a standout majority response. Prince has the most responses, but only by a margin of one.

Here comes the interesting part. For the open-ended, "Tell us why" question, the responses were a bit all over the place. For the sake of simplicity, here's the first half of the responses in Table 8.3.

As you go through and tag or code these responses, what stands out (other than the thought of, "Wow, people are so different, and so interesting?")? Table 8.4 shows the same responses, with highlighting added to point out some of the most intriguing comments.

TABLE 8.3 QUOTES ("VERBATIMS") FROM PARTICIPANTS

Participant	Musician Selected	Comments
Participant 1	Prince	"Let's Go Crazy" is the most fun song ever. It's on my running mix, and that guitar solo at the end, holy cow, the technical expertise! I also like how Prince wrote songs for other singers like Sinead O'Connor and The Bangles.
P2	David Bowie	"Hunky Dory" is the best album of all time, so of course I replied "Bowie" to this question. He has the strongest catalog of any of the musicians listed, with Prince being the runner-up.
P3	Dolly Parton	Dolly donates huge amounts of money to great causes, such as to eradicate childhood poverty. The world needs more people like Dolly.
P4	Dolly Parton	"I Will Always Love You" is the best love song ever. Nothing more to say here!
P5	Aretha Franklin	"Respect" is a feminist anthem, and I get goosebumps when I hear her sing.
P6	Christina Aguilera	Christina can hit high notes like nobody else. Mariah may have a better range, but Christina's songs are better to dance to, and I suppose I like her so much because I am happiest when I'm dancing.
P7	Prince	His lyrics are really poetic—he's really a poet when you get down to it. And his later albums are underrated for their political commentary.
P8	Dolly Parton	I read about how hardworking Dolly is and how she barely sleeps because she's so busy with charity work like childhood literacy, though she doesn't brag about it or publicize it. She's giving back, and I think that's awesome.
P9	Prince	"Purple Rain" is an album I can listen to all the way through, on repeat. He also influenced so many musicians, and I like how he stood up to his record label.
P10	Aretha Franklin	Her voice is angelic and heavenly, a simply outstanding talent.

TABLE 8.4 A FIRST PASS AT HIGHLIGHTING NOTABLE
QUALITATIVE FEEDBACK

Participant	Musician Selected	Comments
Participant 1	Prince	"Let's Go Crazy" is the most fun song ever. It's on my running mix, and that guitar solo at the end, holy cow, the technical expertise. I also like how Prince wrote songs for other singers like Sinead O'Connor and The Bangles.
P2	David Bowie	"Hunky Dory" is the best album of all time, so of course I replied "Bowie" to this question. He has the strongest catalog of any of the musicians listed, with Prince being the runner-up.
P3	Dolly Parton	Dolly donates huge amounts of money to great causes, such as to eradicate childhood poverty. The world needs more people like Dolly.
P4	Dolly Parton	"I Will Always Love You" is the best love song ever. Nothing more to say here!
P5	Aretha Franklin	"Respect" is a feminist anthem, and I get goosebumps when I hear her sing
P6	Christina Aguilera	Christina can hit high notes like nobody else. Mariah may have a better range, but Christina's songs are better to dance to, and I suppose I like her so much because I am happiest when I'm dancing.
P7	Prince	His lyrics are really poetic—he's really a poet when you get down to it. And his later albums are underrated for their political commentary.
P8	Dolly Parton	I read about how hardworking Dolly is and how she barely sleeps because she's so busy with charity work like childhood literacy, though she doesn't brag about it or publicize it. She's giving back and I think that's awesome.
P9	Prince	"Purple Rain" is an album I can listen to all the way through, on repeat. He also influenced so many musicians and I like how he stood up to his record label.
P10	Aretha Franklin	Her voice is angelic and heavenly, a simply outstanding talent.

So what themes are percolating here? Well, there are many—and they are notably different!—but there are themes bubbling up nonetheless. Some comments are about the musician's talent. Some mention the emotions that the music elicits. Several are *not about music at all*, but rather the musician's activities outside of the music world. There appears to be a pattern potentially bubbling up around Dolly's contributions in particular. Hmmm...interesting! Again, this is where your content researcher Spidey Sense ought to be tingling.

If you were the content manager for this music app, you would think hard about what these comments indicate. Several comments focus on charitable work—that's jumping out.

Remember, research is not done in a vacuum. It's worth contemplating these comments in the context of all that's going on in the world right now. Someone like Dolly who is inspiring, selfless, and empathetic is someone who makes other people feel really good. Who doesn't need more of that?

Remember, this is just the first half of the qualitative responses. The responses from the remaining 10 participants still need to be reviewed, of course, but even halfway in, there's something emerging here in these qualitative responses that's worth pondering.

For the first 10 responses, the comment that probably stands out the most is this one: "The world needs more people like Dolly." (Yes, yes it does!) If you were creating content for this music app, you'd be wise to consider these comments around charitable giving. *It's truly fascinating that several participants mentioned charitable and political topics when the research question didn't mention those topics at all!* You have right in front of your face here a topic that's probably worth featuring on the app.

Table 8.5 shows the same qualitative responses, with a new column added for their relevant theme or topic, along with an additional column for the outstanding or most memorable quote.

TABLE 8.5   PARTICIPANT COMMENTS, HIGHLIGHTED AND SORTED INTO THEMES

Participant	Musician Selected	Participant Comments
Participant 1	Prince	"Let's Go Crazy" is the most fun song ever, it's on my running mix, and that guitar solo at the end, holy cow, the technical expertise. I also like how Prince wrote songs for other singers like Sinead O'Connor and The Bangles.
P2	David Bowie	"Hunky Dory" is the best album of all time, so of course I replied 'Bowie' to this question. He has the strongest catalog of any of the musicians listed, with Prince being the runner-up.
P3	Dolly Parton	Dolly donates huge amounts of money to great causes, such as to eradicate childhood poverty. The world needs more people like Dolly.
P4	Dolly Parton	"I Will Always Love You" is the best love song ever. Nothing more to say here!
P5	Aretha Franklin	"Respect" is a feminist anthem, and I get goosebumps when I hear her sing
P6	Christina Aguilera	Christina can hit high notes like nobody else. Mariah may have a better range, but Christina's songs are better to dance to, and I suppose I like her so much because I am happiest when I'm dancing.
P7	Prince	His lyrics are really poetic—he's really a poet when you get down to it. And his later albums are underrated for their political commentary.
P8	Dolly Parton	I read about how hardworking Dolly is and how she barely sleeps because she's so busy with charity work like childhood literacy, though she doesn't brag about it or publicize it. She's giving back and I think that's awesome.
P9	Prince	"Purple Rain" is an album I can listen to all the way through, on repeat. He also influenced so many musicians, and I like how he stood up to his record label.
P10	Aretha Franklin	Her voice is angelic and heavenly, a simply outstanding talent.

Theme or Topic and Subtopic (if Applicable)	Notable Quote or Verbatim
Emotions elicited by music: happiness (fun) Activities participated in while listening to music: running Musician's talent: guitar playing	
Musician's talent: songwriting	
Musician's activities outside music: philanthropy	"The world needs more people like Dolly." —Research participant 2, Female, 35, Chicago, IL, USA
Emotions elicited by music: awe & inspiration	
Emotions elicited by music: happiness Activities participated in while listening to music: dancing	
Musician's talent: songwriting Musician's political commentary	
Musician's activities outside music: philanthropy	
Musician's legacy Musician's activities outside of music: self-advocacy/creative freedom	
Musician's talent: voice/singing	

To sharpen the insights that are slowly coming into focus, pause and take a step back to review the column where you noted the categories or topics for the highlighted info. Are any of the topics similar or even overlapping? Do some topics appear repeatedly? Could those repeated topics be further refined by adding subcategories (or sub-subcategories)?

In Table 8.6, you'll see the collection of responses to the music-app research question, with a distillation of the themes or topics and subtopics, when applicable, in Table 8.7.

It's not so surprising that "talent" appears several times. And it's not super surprising either that participants mentioned that they feel emotional when listening to their favorite singer. "Activities participated in while listening to music" is a pretty intriguing category. Perhaps there is fodder there for some potential new content for your app. But the category that is still standing out the most is "Musician's activities outside of music." This is an unexpected—and really thought-provoking—category that will deserve further discussion, and perhaps further research by you and your team. Yep, here, you have a golden nugget from content research.

TABLE 8.6 LIST OF THEMES AND TOPICS IN QUALITATIVE RESPONSES, INCLUDING SUBCATEGORIES

Activities participated in while listening to music: running
Activities participated in while listening to music: dancing

Emotions elicited by music: awe and inspiration
Emotions elicited by music: happiness (fun)
Emotions elicited by music: happiness

Musician's activities outside music: philanthropy
Musician's activities outside music: self-advocacy/creative freedom

Musician's legacy

Musician's talent: guitar playing
Musician's talent: songwriting
Musician's talent: voice/singing

Musician's political commentary

TABLE 8.7 DISTILLED LIST OF TOPICS, IN ORDER OF
FREQUENCY, WITH FREQUENCY OF SUBTOPICS

Topic and Frequency	Subtopic and Frequency
Activities participated in while listening to music (2)	Running, dancing
Emotions elicited by music (3)	Awe, happiness (2)
Musician's activities outside music (3)	Philanthropy (2), self-advocacy/creative freedom
Musician's legacy	
Musician's talent (4)	Guitar playing, songwriting (2), voice/singing
Musician's political commentary	

This categorization step, or *affinity mapping,* can get quite involved and complex, especially if you've conducted interviews as part of your research, or if you had many, many research participants. As seen in Figure 8.3, you may need or choose to use broad topics to categorize the research highlights.

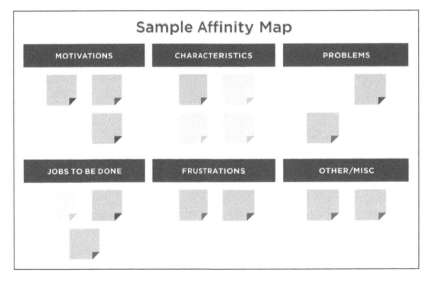

FIGURE 8.3

An affinity map helps organize feedback by showing several broad "umbrella" categories including customer motivations and frustrations.

The hard part is done! You combed through and carefully read the responses, highlighted the standout comments, and then categorized the responses into themes. Now, consider which of the categorizations are most important to you and your stakeholders. This step will help you solidify your next actions for updating your content and will be worthy of flagging or highlighting in your research report.

With this step, you need to know about your content ecosystem and what the current state of your content is. Do you have any content on Dolly's activism? If your app is mostly about music and songs, perhaps her activism isn't an appropriate topic to cover. However, if your app does feature profiles of musicians and their lives beyond their identity as music makers, then your qualitative data has identified a prime content gap to fill. Create some content about Dolly's philanthropy around childhood literacy—your readers are interested in her philanthropy and that of other musicians. What format that content takes is probably worthy of additional research—podcast, video, Q&A-style interview, or something else?

Takeaways from Open-Ended Responses

As a content pro, what can you do with the information gleaned from the qualitative responses? Of course, you still need to evaluate the next 10 responses. Some additional themes may bubble up; some themes that appeared in the first batch of 10 responses may come into focus as being more popular or appearing more frequently than they did in the first batch of 10. Sit with this information and use your judgment about how to summarize it in your research report, and what it means in terms of next steps for you and your content team.

Making It All Work for You

Quantitative research evaluation is usually quick, easy to do, and straightforward. You should have a fairly easy time evaluating quantitative questions and determining what actions you should take with your content to reflect what you've learned, assuming that you have structured your research to minimize bias and align best practices.

Qualitative research evaluation, on the other hand, takes some intense time and effort, although it pays off by potentially identifying your own blind spots. Qualitative research also uncovers major content gaps, reveals fascinating insights about what's going on inside your customers' heads, and, ultimately, empowers you and your team to dramatically clarify and strengthen your content.

SCALING RESEARCH: MANY HANDS MAKE LIGHT WORK

An Interview with Irish Malig,
Senior Experience Design Manager

Irish Malig is a senior experience design manager who has worked at organizations including Meta, Premera Blue Cross, and New York City's Memorial Sloan Kettering Cancer Center. She's a vocal and longtime advocate of content research as a key practice in the creation of exceptionally clear and effective customer experiences.

After more than a decade of working on digital customer experiences, Irish has honed her approach to coordinating research across the product team and has built a reputation for leading teams that are known for their ability to create business impact and influence business decisions.

She structures her research team's operations for efficiency and to scale as easily as possible. She and her team get the partners involved—content strategists, product managers, product designers, and front-end software engineers—to evaluate research responses and note the standout verbatims or quotes, instead of having one or more content strategists evaluating content research results. Then they compare insights and work on affinity mapping as a group. (This can be done online, using Miro, FigJam, or similar tools.)

Her team conducts a high volume of research, reduces bias in the analysis process, uncovers plenty of insights, and most importantly, builds trust across the company by taking a "more-hands-make-light-work approach." She's found that content research "greatly reduces risk for the business." When content-focused research debunks a PM's hypothesis and flags where more clarity is needed, Irish says, "We dramatically change the course of that project. I can say with confidence, 'I stopped you from wasting two months of design and dev time!'"

She also encourages her team to implement content research insights as soon as possible. And by as soon as possible, that often means before the research report is completed. For example, if a front-end dev in her organization is watching a recording of a research session with a customer, "They'll jump into the designs and be annotating and revving [revising] the designs to reflect the recommendations that they just heard [from the customer], right then and there."

Share Results from Content Research Studies

When you share the results of your research, the thorough and thoughtful planning that you put into setting up your study pays off. You used a well-organized, concise test plan. You also collaborated with your core group of stakeholders and had them review and provide input on the test plan before the study started. You wisely scoped your study to focus on a precise audience with specific content needs. Congratulations, you're already two-thirds done! Now, you can take that test plan and quickly leverage it to create a clear, easy-to-follow, shareable document of the research results—a record of your work that will provide value to your team and company long after the research is wrapped up.

Before getting into the frameworks and formats you can use to share your research results for the greatest impact, there's a topic that absolutely must be covered first: tact.

The Need for Tact When Sharing Results

No matter how well-organized you were with the planning, setup, and running of your study, the essential thing to keep in mind as you share research results is tact. It's of paramount importance to be tactful, kind, sensitive, and diplomatic when you share research results, especially if your research evaluates and gathers feedback on content that's currently live and already being seen by your customers. In the case where you're using research to iterate and improve upon content that's already customer-facing, the odds are good that the content was originally created by one or several of your current colleagues. (If you have a small content team or you're a brave content team of one, it could very well be that you yourself were the person who created the original content.)

No matter who created the content that's being evaluated by research, be careful not to insult the content that's being iterated—and by extension, you or your colleagues. Frame the feedback from a point of view of continual improvement (*kaizen*) and a growth mindset.

When working in content, you need to develop a thick skin and embrace feedback, but it can take years to do so. It doesn't hurt to keep this quote from the Dalai Lama posted by your desk: "Be kind. It is always possible."

Tact is essential when sharing content research results, for three reasons:

1. **To avoid offending the person or people who created the content you're evaluating:** If you've ever had someone inform you that they feel your content isn't engaging or strong enough, you know how this can sting. That kind of feedback is something you can recall vividly many years after the fact, and it chews away at your self-esteem.

   Whether or not the original creator of the content you're evaluating is still at your company, it's just plain kind to use kid gloves and a constructive approach when sharing research results. It's smart to emphasize that the feedback you're sharing is coming from your customers or potential customers, and not you, the content researcher, or other stakeholders themselves. This approach provides a protective layer of objectivity.

   If you do know which content writer, content strategist, or content designer is responsible for the original content that you're evaluating, it's wise to share results with them—and them only—in private, *before* sharing it out to any additional colleagues. If there's one thing that's more painful than having your writing critiqued, it's being unpleasantly surprised by seeing research results for the first time when they're being shared in a forum of peers and superiors. Ouch!

   It may be that the person who created the original content is not a content pro, due to staffing or other issues. In this case, remember that not all of your coworkers love writing and creating content the way you do. Writing is challenging for many people. Trust that whoever created the content gave it their best shot.

2. **To avoid a critical, condescending tone with your content peers and stakeholders:** Condescension has the potential to overshadow you and your content research program with a reputation for being critical, instead of open-minded, collaborative, kind, and helpful. Think good vibes. Constructive feedback. Helpfulness. All those good things! Content teams are often understaffed, underappreciated, and struggling to get a seat at the table. Please don't undermine any social capital, trust, and respect that you and your team have built by coming across brashly.

3. **To protect you and your content team:** Tact is also incredibly important when you're sharing research insights and results in front of any coworkers, especially those who hold rank above you in the organization chart—say, your manager's manager. The last thing you want to do in that situation is make that leader think that you or your team have allowed subpar content to be published and seen by customers, for any amount of time.

It's important to frame content research as a customer-centric methodology that's helping the content and product team to always be improving. It would be a career-limiting move to describe any customer-facing content as not very engaging, not very effective, or not very clear. (And certainly not as "awful" or "dreadful" or anything blatantly critical like that.)

The truth is, perhaps your research uncovered how participants did indeed tell you they felt the content in question was hard to understand, dull, or worse. Be very diplomatic and judicious when sharing any "verbatims" or direct quotes from participants. This is especially important when you're working with an audience that may not be well versed in the complexity and challenges of creating effective, strong content. (Yep, I'm referring to the C-suite and senior leaders in your company, as well as others on your product team, from product design to engineering.)

Again, it's key to approach content research from a growth mindset or an "always-be-learning" point of view—not a tearing-apart-the-work-of-your-peers point of view. You didn't know as much about your content (or your customers' thoughts or opinions) before you invested the time and effort in research. Content research helps you improve your team's work iteratively, together, for the sake of the user experience. You can accomplish that without crushing anyone's self-esteem.

So, how exactly do you put tactfulness into practice? Think about how you would want your content to be evaluated by others. Be sensitive whenever sharing results, whether you know who specifically was involved in its creation.

Here are some dos and don'ts.

Steps to Avoid When Sharing Results

It's important to avoid some potential pitfalls when sharing research results. Do take care to be as objective as you can with the adjectives and descriptions you choose to use in your research summaries and report-outs. Remember, content research can be thought of as a form of feedback, and receiving feedback is often challenging for most people. Be careful to frame the research results as constructive criticism that's been shared by your customers (and not directly coming from you or your content team). The feedback is being used to improve your customer experience, for the sake of improving business performance. Be careful not to frame the feedback as a direct criticism of anyone's writing style or content skills in general. The following tips can help you and your stakeholders proceed with sensitivity as content research results are shared.

What *Not* to Do When Sharing Research Results

At all costs, avoid using judgmental descriptions in your research reports. Here's a list of descriptions to avoid when describing content. (Warning: Some of the examples listed here include language that's flat-out insensitive, especially to people with mental illness or other disabilities.)

- "Good" or "bad" content
- "Best" or "worst" content
- "Winner" or "loser" content
- Crazy
- Lame
- Dumb
- Stupid
- Pathetic
- Shocking
- Under-performing
- Sub-optimal

This point was mentioned earlier, and it bears repeating: Content research is *not* a battle, or a competition! Rather, it's a structured way for you and your team to understand your customers or audience and their challenges more fully, and to bring clarity to which words and phrases work, in which scenarios, and why. There's simply no need to color any content results with a competitive tone or even "war"-like point of view. Nor is there any place here for hurling insults (not to mention they violate all guidelines of sensitive and inclusive language). You wouldn't dare use any of these belittling adjectives to describe your customers or your colleagues. By all means, please avoid using loaded adjectives to describe the content.

For quantitative questions, here are some descriptions to use when one content option is preferred by a strong majority of participants (instead of using judgmental descriptions like "best" or "winner"):

- Content **that resonates with** research participants more or less than other options.
- Content **that appealed to** research participants more or less than other options.

Using tactfulness and framing your test results in a constructive way is also key to creating and sustaining a healthy content research program that has a positive, uplifting vibe to it—and therefore continually attracts more participants from across your team and around your company. Maintaining a positive vibe will also help ensure that your research program continues to get support from the managers whose budgets pay for your research tools and staff.

Three Steps for Sharing Research Results

Now that tact is on the top of your mind, it's now time to communicate the results of your research study. As tempting as it may be to cut to the chase and immediately fire up your content management system to update the content so that it reflects the insights you uncovered (if, indeed, your insights point you in such a direction), do yourself and your team the immense favor of *squashing that temptation*. Instead, simply start with documenting the results. Then go a step further and save the results on a shared drive or wiki or in whatever format will render the research discoverable—and ideally searchable—in the future by your content team, product managers, software developers, product designers, and senior leadership. These steps are outlined in more detail below, in Figure 9.1.

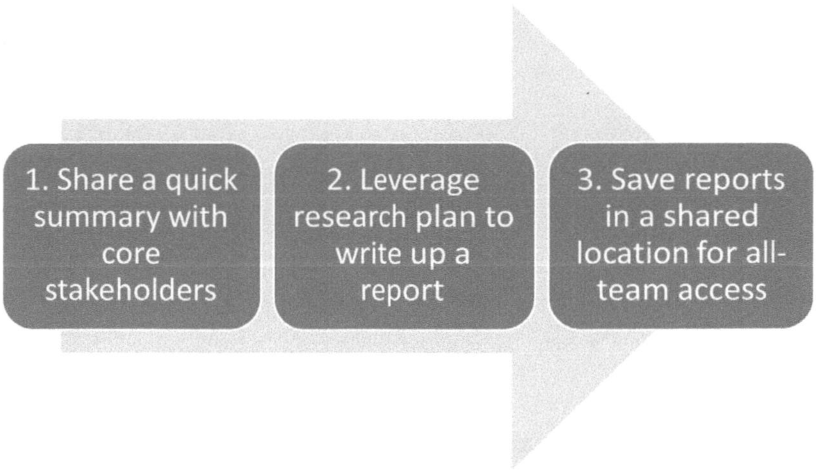

| 1. Share a quick summary with core stakeholders | 2. Leverage research plan to write up a report | 3. Save reports in a shared location for all-team access |

FIGURE 9.1

The three basic steps of sharing research results to help ensure that colleagues can easily find and refer to the results, now and in the future.

Step 1: Share a Quick Summary with Core Stakeholders

When you've completed a study and have distilled the core quantitative *and* qualitative insights, start with sending a quick email, group chat message, or other notification to your core research team (typically, your product manager, visual designer, user researcher, and possibly your software developer). Why a quick heads-up message first? Chances are excellent that your feature team has been reaching out to you as the study's been in progress, eager to hear results. Or maybe you're the one who is eager to inform your team of what's been learned so that you can make the case to your team that content updates are needed. Or, in the case when research indicates that *no* updates are needed, it's still best to let your team know of this discovery sooner, rather than later. Whatever the results, be transparent, be communicative, and update your team as soon as possible.

Figure 9.2 is a sample email communication for sharing the results of a terminology research study.

Sample Email Template for Sharing Test Results

Subject line:

Start each email subject line with "Content research results:" followed by a short but informative synopsis. For example, *Content research results: Terminology preferences of new health-plan customers in the new-customer onboarding experience.*

Body of email:

You may simply copy and paste this key info directly from your study plan. You can also include a link or attachment to the full study plan for anyone who wants to read the whole enchilada.

Date(s) that research study was conducted & location (if conducted in person):

Number of participants:

Focus audience:

Persona, geography, demographics, etc.; info from any screener question if it's essential information.

Why the study was conducted:

Provide a short summary of what specific information you and your feature team sought or what questions you were seeking to answer.

Qualitative data highlights:

Include a representative customer verbatim statement/quote or two. Note: Listing qualitative results first can help drive home the importance of qualitative data across your team. It can also help stem the tendency of teams to share only the quantitative results, especially when only sharing donut charts or bar charts without the oh-so-essential underlying details that explain the user's thinking behind the quantitative data.

Quantitative data summary:

A one-sentence distillation. Include any graphic or chart if available—bar chart, donut graphic, pie chart, etc.

Link to your study/test plan document:

Any additional links that you can share for those people who want to deep-dive into the granular details:

This may include links to participant videos, the spreadsheet of qualitative feedback/verbatims, and direct links to the study if you used a platform like dscout or UserZoom.

Next steps:

This part is essential! Based on any golden nuggets or other insights, provide clear next steps for improving content. In the case when the research is inconclusive, provide suggestions for further research.

Example: Based on this research, we recommend creating a series of onboarding email notifications for new customers sent a week apart.

FIGURE 9.2

Anatomy of a sample email that summarizes the results of a content research study, including study targeting, insights, and follow-up actions, with recommendations called out.

THE NEED TO BE CONSISTENT WITH RESEARCH COMMUNICATIONS

When you and other content researchers notify stakeholders and colleagues about study results, it's helpful to be consistent with the chat message or email subject-line format.

And, whenever possible, be consistent with the format of the research summary and insights themselves. Just as it's a huge time-saver to use a template when creating your research study, it's also helpful to templatize or at least make your communications about research results as consistent as possible. One simple way to create this coherence is to start each message with "**Content research insights from [the topic of your study]**." You and your coworkers will find this consistency to be a strategic help as you ramp up your research program:

- The clarity of the subject lines will help grab attention and pique the interest of your colleagues, encouraging them to read the results.

- As you grow and scale your research program and run additional studies, and as more people across your department or organization start participating in content research, this consistency will also help convey that research is being conducted regularly. A regular "drumbeat" of research helps to convey that content research is important and worthy of attention.

- Lastly, this consistency will also help you and your colleagues search for and find research results more easily in the future, when you search in email or filter chat messages (such as when you conduct subsequent research studies).

AVOID SHARING SNEAK PEEKS!

When your team starts creating a "drumbeat" of content research, word will spread around your company about your efforts. As word spreads, anticipation grows. As anticipation grows, you'll find that people will find it more and more difficult to wait for any research-summary emails or messages, let alone full reports. Instead, you'll inevitably hear from people—stakeholders involved in the study, people not involved in the study, and sometimes even people you don't even know and have never worked with yet. They'll ask you for a "sneak peek" at the in-progress study results.

A word to the wise: Do not share preliminary or partially complete study results. For example, if you're running a study with 20 participants, don't cave to the temptation to share preliminary results with anyone when you have responses from only 10 or 15 people. This holds true even if it looks like there's going to be a clear preference or majority response. It also holds true even if your team is desperately awaiting insights for a soon-to-launch campaign, website, or what have you. This absolutely holds true if someone in your C-suite reaches out and asks for a "quick look."

Here's an analogy: The Seattle professional soccer team, the Sounders, show short videos for halftime entertainment. At each game, there's a cartoon-y video played for the whole stadium on the giant LED screens above the home team goal post that depicts a race of three speedboats. There's a yellow, red, and blue boat. People in the crowd cheer for their favorite color. It's cheesy. Inevitably, one boat is significantly ahead for most of the race, and another comes from behind with just a couple of seconds left for a surprise finish. (Truly, it's curious how a stadium full of soccer fans will cheer on a trio of cartoon boats.)

Content research, like all user research, is similarly unpredictable. If you're running a study with 20 participants, and you share initial results that include input from only 10 participants, you may come to rue the day that you shared it. Colleagues will talk (or email or message or chat) and may jump to conclusions and even make content changes based on only a limited amount of study results—just like one of the cheesy Sounders boats may come from behind when it initially appeared that it didn't have a prayer. Then you face a communication challenge to "correct" your colleagues about the accuracy of the results. And all this serves to erode the credibility of you and your content research program.

Step 2: Leverage the Research Plan for Your Report

You quickly shared with your core team a short and sweet summary of the study results. Now, it's time to write that report. No procrastinating! Tackle this while it's fresh in your mind.

Take the test plan and add a few short notes about your quantitative or qualitative results. Note any golden nuggets or especially intriguing insights or discoveries. If you used a research platform like dscout or UserZoom, include a link to the study itself. If you have any key video clips or notable screen grabs of participants' facial expressions, or verbatim quotes, include them, too.

Figure 9.3 shows a graphic that conveys how to simplify your content research process: take the content research template, turn it into a research plan, and then leverage that plan to transform it into a full research report.

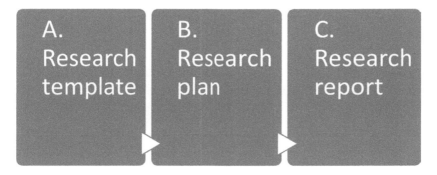

FIGURE 9.3
Three phases of content research documentation: a content research plan test template, a research plan created from that same template, and a research report created from the research plan.

The key is to be speedy here, so as not to lose momentum or let the zing of energy you feel from running a research study fade before you go about the task of documenting the results. If you are not yet a convert to the "Pomodoro Technique," the writing of your report is a fine time to get to know it. Named after a tomato-shaped egg timer, the Pomodoro Technique is the practice of setting a timer to help limit the amount of time you spend on a specific task. Set a timer for 20 or 30 minutes and try to write up your report as quickly but accurately as you can.

Be sure to give credit to any feature team members who helped develop the study plan or gave input. For the format, Word or Google Docs works fine; if you work in a particularly visual industry or your office culture dictates it, you may want to use PowerPoint or Google Slides. (To enable your colleagues to search easily for specific information or details within reports, you'll need to stick to using documents rather than presentations.)

Save this report somewhere safe. When you're just starting your content research program, it's very helpful for others to be able to access examples of reports, so a shared folder is key. Again, be consistent with your naming convention, for the sake of supporting the growth of your research program. Save the report somewhere that's ideally accessible to your entire team, or even better, your whole organization. Make sure that it's available more broadly than just the immediate feature crew who were involved in the development and running of the study.

If you're super organized (and perhaps caffeinated), you can attach this report to the summary communication that's outlined in "Step 1: Share a Quick Summary with Core Stakeholders." However, if your team is enthusiastically awaiting the results, chances are good you'll need to share the summary communication without the full report. (Don't worry, the report will be put to good use—more on that in a bit.)

While you may be a bit wiped out from all the activity involved in planning, reviewing, creating, and running your study, it's crucial to get the report written while the information is fresh and vivid in your mind. The longer you wait to start or complete the report, the more painful it will be to do the actual writing—and likely the longer it will take. Do yourself a favor and harness the energy you felt when you spotted an insight or discovered something key to your customer experience and think about the peers and colleagues who can benefit from that information, and just do it. (It may be tempting to outsource this writing to an intern or coworker, but by the time you provide direction and guidance, you probably could have already written it up yourself.)

Step 3: Save Research Reports for Sharing and Future Reference

Until you get your content research practice off the ground, keep all your reports in a safe, accessible place. A folder on a shared drive will do. The key is to be open and transparent about the research. Aim to make it as easy as possible for others to access the reports and to be able to share them with others. (Set the permissions level of the reports so that only you and the core team of researchers will be able to edit them.)

As you and your team conduct additional research studies in the future, you'll be glad that you don't have to go hunting on your own laptop or reach out to your fellow content teammates to ask them to dig up and share their research results. Storing all reports in a single place will also be key as your team's research practice grows, and as new people join your team or organization and want to view results of past studies. Finally, it'll be key for when you have a large enough body of research such that you're able to conduct meta-analysis to evaluate your research for broader patterns, like variations in audience-specific insights.

If you're fortunate to have resources to eventually compile a searchable repository or wiki of research studies, having all studies in one spot will again serve you well. And again, keeping organized will build trust and communicate that your team is serious about the value of content research, and that it's something that others across your department and company ought to be paying attention to, too.

PRO TIP **STANDARDIZING THE FILE-NAMING FOR REPORTS**

Keeping your research reports in a single, safe, shared folder or drive is smart. It's also smart to create a file-naming convention for the reports, so that it's as easy as possible to find reports within that shared folder or drive. If you consistently use a clear, informative format for the file names, you're bound to thank yourself rather than curse yourself later on. Encourage your team to avoid vague file names like, "Product Z content research." Instead discuss, agree upon, and adopt a file-naming structure such as, "Content research_Audience_Topic_Initials of content lead_YYYYMMDD."

For example, a file name like "Content research_entrepreneurs_ Product XYZ messaging_SK_20240822" is informative and helpful. "Entrepreneurs" is the audience, "Product XYZ" is the product name, "Product XYZ messaging" is the focus of the research, "SK" are the initials of the lead researcher, and "20240822" is the date the study took place. This sort of descriptive file name is also much more likely to be useful to others across your company, compared to a nondescript, undated file name like "Product Z content research."

Frameworks for Communicating Content Research Insights

No matter whether you're a content research team of one, or if you have a dozen or more content pros participating in your research program, it's smart to formalize and eventually operationalize the way you share results. You put hard work into the research, so make sure that you widely share the knowledge you've uncovered with coworkers who can benefit from this hard-won information.

While your insights will arguably be the most valuable for the content team, don't keep this information confined to content pros. Spread the word as broadly as you can to help your whole team, organization, or even your entire company understand just how much words matter. Depending on the maturity level of your research program and your content staffing, take your pick of the following communication frameworks for getting the word out.

The "Crawl" Approach: Emailing or Sharing One or More Reports

Just like it's quick and easy to email your hot-off-the-presses research insights to your core feature team, it's easy to share your reports broadly via email or chat notifications. This simple style of sharing isn't as attention-getting or as polished as the more operationalized and formalized methods of sharing that are outlined in the following "Walk" and "Run" sections. But you need to start somewhere. It's better to communicate, period, even if it's in a very basic way, than it is to hoard the insights and not share the research findings at all.

If there are key stakeholders or colleagues whom you're particularly hoping will see your work, use email as your mode of communication, and use the "Read/Receipt" feature so that you can clearly see who has viewed it. You can be bold and wait a couple of days after sending such a message and directly follow up with any extra-important stakeholders to ask whether they have any questions (and perhaps to verify that content updates are in progress, if you're able to quickly get started on that work).

The "Walk" Approach: The Research-Roundup Newsletter

When I worked at Expedia, the amount of usability, and, specifically, content-focused research that was being conducted was enormous. Truly, it was hard for anyone to keep up with all the research (as well as A/B experimentation) that was going on. The lead researcher on the UX team decided to help everyone in the organization be better able to keep tabs on the research by emailing a distilled, easy-to-scan recap of research insights each week. Sometimes it would contain info about five research studies; sometimes it was 20.

Of all the hundreds of emails I received every week, this summary email was the one I read from start to finish—as did plenty of colleagues. The best part about this email was that it created anticipation and energy, because it got people talking. You'd hear coworkers in the elevator talking about some specific study results, or chatting about a study before a meeting began, or even while in line at the espresso shop in our building. These regular emails helped establish the research program and created a regular, reliable "drumbeat" of information sharing. The info was spread even further via word of mouth. I made a point of saving these emails for future reference.

This sort of anticipation around information sharing generates a lot of attention, and it also builds trust and respect for the research program. If you can count on receiving an email or message that arrives on a regular schedule and it packs plenty of intriguing, valuable customer insights, you're going to be looking for it—and the person sending this information is going to gain a great deal of respect and influence within your product organization.

The "Run" Approach: Create a Formalized Sharing Platform or Wiki

At Amazon, the huge volume of content research conducted on website promotions and email campaigns created an avalanche of content insights. These insights were continually added to a wiki that was shared across the content leads. At Microsoft, usability research of all kinds, including content research, is shared on a SharePoint site called *HITS—Human Insights Team Studio*. The HITS platform is a wonderful way to make research reports accessible to the whole company—more than 100,000 employees. It's easily searchable by topic, product or feature, author, audience, and type of research.

This searching is enabled by meta tags that are added to each report. My team instituted a tag specifically for "content research," so it was easy to filter the thousands of studies and comb through those that focused directly on content. One additional helpful feature of the HITS report platform is that it indicates how many colleagues have read each research report. When you can see that hundreds of coworkers have read one of your research reports, it's quite gratifying.

Another plus to implementing a SharePoint or other wiki-style platform is that you can easily cite and link to other studies that have been saved to that same platform. This approach helps visually highlight the importance of content research. For example, when one research study cites a dozen or more other content research reports, or, on the other hand, *is cited by* many research reports conducted after its publication, it clearly demonstrates the durability and value of the research (like citations in academia).

The "Fly" Approach: Meta-Analysis

Once you have a collection of reports—maybe a dozen or so, although more is helpful—you can start looking for what trends and patterns or intriguing stories emerge when you look at the overall body of research. In other words, you can conduct a meta-analysis.

For example, after completing several content research studies that focused on a specific audience, can you pinpoint trends in your insights, or even use your research as the foundation for developing validated audience personas? If you've run several studies that evaluated terminology preferences for a specific product or audience, can you now confidently make statements about the effectiveness of the overall voice and tone of these words and terms, or make other bird's-eye evaluations?

Table 9.1 summarizes the "crawl-walk-run-fly" approaches for sharing content research results.

TABLE 9.1 CRAWL-WALK-RUN-FLY FRAMEWORK FOR
COMMUNICATING CONTENT RESEARCH RESULTS

Content Research Program Maturity Level	Share-Out Format and Audience	Pros and Cons
1: Crawl	**Format** One-off email including a link to a single research report or links to a small handful of research reports **Recipients** Your core feature team, plus other individuals or teams who will benefit from seeing the insights	**Pros** It's quick! This is an effective way to share results with your team beyond the immediate feature crew that was involved in the research study itself. If you use email and its "read/receipt" feature, you can track how many people open and read the email, making it easier to know who is aware of the results. **Cons** Although it's not particularly polished, this approach is better than not communicating results at all. However, it doesn't generate as much trust and confidence in your research program as it could by using more operationalized formats. It may ruffle feathers and unintentionally offend colleagues if you omit them from the recipient list.

continues

TABLE 9.1 continued

Content Research Program Maturity Level	Share-Out Format and Audience	Pros and Cons
2: Walk	**The Roundup** A regular newsletter that includes links to multiple reports from all the most recently completed content research studies; it could be a weekly, monthly or quarterly collection, depending on the volume of research your team is completing, and how urgently and quickly you need to share insights. Recipients Your broader team or even your whole organization (so long as recipients have permissions to read the reports you share)	Pros It creates anticipation and water-cooler buzz. The newsletter can be easily created by leveraging your research study plan. Collections of research studies can be grouped in one beefy, insights-rich email, which improves the odds that it will be read widely, and that people will anticipate its arrival each month. It's easy to share or forward to others, which means it can help spread the word and broaden the influence of your research program. At the end of a quarter, or every six months or so, it's a great idea to take all the newsletters and compile them into one mega collection of juicy content research goodness. This collection of research can reinforce the importance and value of your research practice, especially to senior leadership. (These collections can even be used to help with recruiting new hires to your content team!) You can create an accompanying email alias, specifically for content research, and encourage people around your company to contact that alias to ask questions about the research studies to date. Cons When you round up research results this way, they're not in a searchable, filterable format. That can make it hard for new hires to find all the research info they need to familiarize themselves with the business, the audiences/customers, and so on.

continues

TABLE 9.1 continued

Content Research Program Maturity Level	Share-Out Format and Audience	Pros and Cons
3: Run	Formalized Sharing System or Repository (like a SharePoint, Dropbox folder, or a customized wiki)	**Pros** The "run" approach creates a one-stop shop for research and insights. You can track how many people have accessed an individual report, which can be useful proof of the effectiveness and importance of your research program. This kind of formalized research can also help drum up attention and leadership support for a continued budget for your research platform or survey tools and other resources for your research program. Reports are shareable and can be easily searchable. Reports can be easily filtered if you use meta data (for audience, persona, product, type of research—for example, content research—and so on). **Cons** You may need technical support to get a sophisticated wiki or repository set up. Reports may be "out of sight, out of mind." At the start, you may need to frequently promote the shared site to encourage colleagues to access it and read the reports. (To get around this, you can make a habit of sharing research highlights at every monthly business review meeting, quarterly all-hands meetings, and other regularly scheduled meetings that focus on business impact.)

continues

TABLE 9.1 continued

Content Research Program Maturity Level	Share-Out Format and Audience	Pros and Cons
4: Fly	Meta-Analysis of Content Research, distilled by persona/audience, product/feature, or business goals, to evaluate for patterns and trends	Pros A meta-analysis finds trends and patterns in research studies that otherwise would go unnoticed. A meta-analysis can function as a content audit of sorts, highlighting if your team is focusing too intensely on one persona or audience at the expense of another. In this way, it can help bring balance to your content research efforts (and content strategy). If your team is interested, you can leverage the meta-analysis into a formal, published industry research report. Cons It requires a solid body of research to be effective. It can be time-consuming! It takes focus and patience to create a meta-analysis (but the time invested is well worth it).

(Beyond conducting meta-analyses, some content teams take a body of content research—the past business quarter, or six months' or a year's worth—and collect it into a handy report, summarizing the number of research topics, audiences or customer segments, and research formats. This is yet another great way to generate attention and respect for your content research program.)

Customize Research Insights to Fit the Audience

Sharing your thorough test research plans and in-depth study reports with your feature team makes sense. However, when you're communicating with senior leaders who are triple-booked all day long and want and need just the essential facts about how your content team is driving business results, you need to distill your research into a very easy-to-grasp, executive-summary format.

Executive summary-style snapshots of your research studies, such as Figure 9.4, are an effective way to let your colleagues and leadership team quickly understand the scope and methodology of your research. They also help executives to easily understand the results—what types of insights are bubbling up?

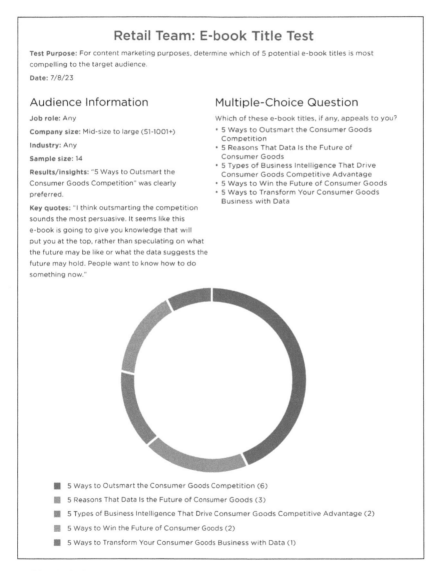

Retail Team: E-book Title Test

Test Purpose: For content marketing purposes, determine which of 5 potential e-book titles is most compelling to the target audience.

Date: 7/8/23

Audience Information

Job role: Any

Company size: Mid-size to large (51-1001+)

Industry: Any

Sample size: 14

Results/insights: "5 Ways to Outsmart the Consumer Goods Competition" was clearly preferred.

Key quotes: "I think outsmarting the competition sounds the most persuasive. It seems like this e-book is going to give you knowledge that will put you at the top, rather than speculating on what the future may be like or what the data suggests the future may hold. People want to know how to do something now."

Multiple-Choice Question

Which of these e-book titles, if any, appeals to you?

- 5 Ways to Outsmart the Consumer Goods Competition
- 5 Reasons That Data Is the Future of Consumer Goods
- 5 Types of Business Intelligence That Drive Consumer Goods Competitive Advantage
- 5 Ways to Win the Future of Consumer Goods
- 5 Ways to Transform Your Consumer Goods Business with Data

- 5 Ways to Outsmart the Consumer Goods Competition (6)
- 5 Reasons That Data Is the Future of Consumer Goods (3)
- 5 Types of Business Intelligence That Drive Consumer Goods Competitive Advantage (2)
- 5 Ways to Win the Future of Consumer Goods (2)
- 5 Ways to Transform Your Consumer Goods Business with Data (1)

FIGURE 9.4

This example of an executive summary for a content-marketing study has an easy-to-read format and provides a brief recap of the test goals, audience targeting, and research results.

These executive summaries can also be easily compiled into quarterly or annual collections that are easy to group by audience, product, or topic.

Presenting "Before-and-After" Versions of Content

If you're able to include it in your executive summaries, it's very powerful to include the updated and improved version of content in your share-out report. Including side-by-side examples of the "before-and-after" versions of your work shows the real power of content research (see Figure 9.5). Before-and-after storytelling also appeals to senior managers and leaders who are overloaded with information all day long and prefer (or need!) communications in condensed, summary-style news bites.

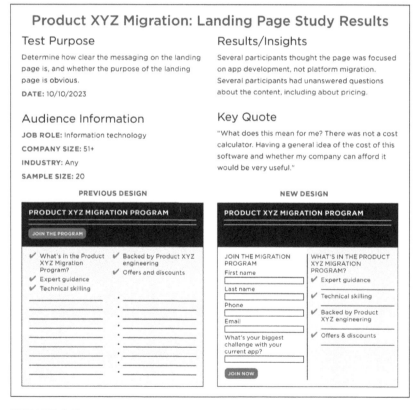

FIGURE 9.5

An executive summary presentation slide that shows "before-and-after" examples of content. Content was iterated to reflect the insights discovered from content research.

The Content Research Open House or "Science Fair"

When you've spread the word about the goodness of content research, have a small group of content pros conducting content research who have collectively been at it for a few months and therefore have an intriguing body of research work to share, you can and should spread the word far and wide. A fun way to do so is to hold an "open house" or "science fair" event to show off your hard work and fascinating insights in a way that will help your research practice thrive even more strongly, whether you're able to be physically in your offices or need to run such an event remotely and online only.

How to do it: Ask each content researcher to prepare to display their most insightful study results—and invite *everyone* in your group or organization to come take a look.

The in-person way: Remember science fairs in elementary school, with students standing next to their big poster boards, enthusiastically sharing what they learned and happy to answer questions about their experiments? Take that concept and run with it, adapting it for your content research results. Setting up tables with old-school printouts of the test summaries and displaying them on poster board or studio boards (like the ones often used by ad agencies for creative reviews) is a fun way to strike up conversations with your broader organization and showcase the hard work you've done to uncover all your golden nuggets of insights and customer and content knowledge.

The digital way: Have the content pros create a FigJam, Miro, or Mural board or presentation slide that summarizes what they learned and how. Schedule a video call, with breakout rooms—one for each content research lead. Ask each breakout group to spend 5 or 10 minutes discussing the study structure, insights, and impact of the content examples. If you have time, you can then switch up the open house attendees' breakout rooms, so attendees are able to see additional content examples.

Making your science fair an annual event is a great way to keep your team's energy level high and maintain enthusiasm and support for content research efforts.

Making It All Work for You

If you're investing the time and effort into conducting content research, your work doesn't stop once you've identified the quantitative and qualitative insights from your study. You need to devote a bit of additional time and effort into writing your reports, saving them in ideally an operationalized way, and sharing your research insights across your product team and company. This process is made easier if you leverage your research study template and already existing research study plan to help you create your summary report quickly.

Do what you need to do to carve out the time needed for this reporting work and communication, and don't procrastinate.

Chances are, it will be worth it for you and your product team to devote a certain number of times each month or sprint to the valuable practice of conducting content research. If this means taking some other work off your plate to help prevent overtime, late nights, or weekend work, do it! (If it means you and your content team stop holding "office hours" for quick-turn content help on projects that don't have a dedicated content pro assigned to them, go for it.) The high return on investment of content research should help you and your content team make the case for regularly earmarking time to uncover content insights, because these insights make your products and features more engaging to customers and easier to use, which translates into a more successful business and stronger bottom line.

CHAPTER 10

More Content Research Techniques

The quantitative and qualitative content research techniques outlined so far can help you solve probably 90 percent of your content research needs. In this chapter, you'll learn about additional content research tools and evaluation frameworks, and other techniques worth exploring:

- Heat maps and gaze plots
- Five-second tests
- Competitive analysis
- SUS and SUPR-Q
- Social media research
- Moderated interviews

Heat Maps and Gaze Plots

When you look at a web page or an app, your eyes scan the content quickly, darting up and down, left and right, and everywhere in between, dwelling for milliseconds on different sections of the content until something catches your eye, or you spot the information you're interested in. Maybe you're looking for a hamburger menu, a link or button, or a navigation list. When you find what you're looking for or something that's interesting to you, your eyes dwell on that specific spot for longer than the areas that are not as interesting, relevant, or attention-grabbing to you. You also click (if you're using a mouse) or tap (if you're using a mobile or touchscreen device) to select one or more links, buttons, or sections of the page.

Each spot where your eyes dwell on the page or where you click, tap, or select the content can be quantified using heat maps. Heat maps are color-coded representations of the specific spots where an audience's eyeballs are focusing and where they're clicking with their mouse or tapping, on a page-by-page or screen-by-screen basis. Heat maps are enabled by special software and sometimes by special equipment called *eye-tracking hardware or software.*

For example, say you have a new puppy and there's a ton of puppy fur suddenly flying around your house. You recall that every dog owner you know seems to own a Swiffer, to clean up the fur. You go visit the Swiffer website (seen in Figure 10.1), and at the top of the page, you quickly scan the navigation headers from left to right, very briefly. (The heat map picks up where your gaze lingers on the site.) Then your attention moves lower on the page, and you look over a description about how Swiffers work. (Ditto: The heat map notes that this is an area of the website that captured your attention.) You decide you want to buy some Swiffers online and look around the website for a "Buy now" button, and when you spot it, you select it. (The heat map notes that the call-to-action button is the last area your eyes focused on. On this specific web page, and because you clicked on it, it's noted as a "warm" area on the heat map.)

A heat map is a useful way to quantify the multitude of mini-decisions and starts-and-stops you made as you searched around and absorbed the information on the web page.

FIGURE 10.1
The Swiffer website content caters to customers with shedding pets.

How Heat Maps Work

Heat map software puts an "overlay" onto your web page or app. Heat maps typically use small, colored circles to denote the specific spots where the audience member's eyes are looking (this typically requires eye-tracking hardware), or where the audience is interacting with the content (such as by hovering a mouse over a specific part of a page).

Heat maps often use green- and blue-colored circles to represent the less-popular or less-engaging dwell spots; yellow to denote a greater level of attention from the audience; and red represents the hot-hot-hot, most frequently and longest dwelled-upon areas. Heat maps can be used to measure where an individual person is looking, or they can aggregate data from dozens, hundreds, or thousands of visitors to your website or app. Fascinating? Yes, absolutely!

Some apps that enable heat mapping include Hotjar, Crazy Egg, and Qualtrics. A sample heat map created with the Hotjar app is shown in Figure 10.2.

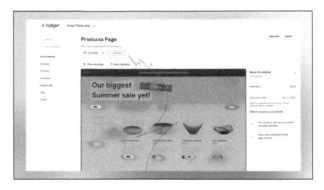

Just like the donut charts and bar charts that are the outputs from quantitative content research, heat maps are data visualizations that quickly tell a story and render them into what can be called a "crowd pleaser" research tool. If you take a heat map and share it with a colleague, you can pretty much bet that they're going to share it with others. And you can certainly bet that they're going to ask you for more heat maps in the future.

A slightly different type of heat map is called a *gaze plot*. Gaze plots take the quantification of heat maps a step further by denoting *how many* people looked at which specific spot on the page. Gaze plots can also quantify *the order in which* customers navigated through your content—that is, they show which elements they looked at or engaged with first, second, third, and so on. For this reason, they're sometimes called *click tales*—as they tell a story by revealing where your audience is dwelling on your page or app screen.

For example, in the Hotjar web page shown in Figure 10.2, each green circle represents where a user's mouse hovered or clicked. The bigger the circles, the greater the audience's interest with that specific spot on the page. Again, like heat maps, this information is intriguing, but you won't find out *why* someone is engaging with your content until you use additional content testing.

Interpreting Heat Maps and Gaze Plots

Heat maps are intriguing because they tell you *where* the eyes of your audience or customers are dwelling. What heat maps don't tell you, however, is *why* the hotspots on your app or web page are capturing the attention of your audience. For this reason, it's helpful to combine heat map data with qualitative research, such as preference or action-ability studies, to uncover the "why."

Sometimes, heat maps and gaze plots can be used for gap analysis. However, you and your team may find them frustrating, particularly when you notice that customers or your audience are unexpectedly *not* clicking or engaging with call-to-action buttons, navigation labels, or other sections of your app or website. Often (though not always), when your customers are not engaging with your content, it's because it's unclear or otherwise off-putting. Use any "cold" areas of your heat maps or gaze plots as data points to investigate further into why customers aren't clicking or tapping where you expect them to. (Hint: It might be due to distracting banners or promotional messaging or other prominent design elements on the page. This is where collaborating closely with the marketing team, product managers, and other colleagues is important!)

Conversely, if you see that your audience is engaging to a great degree with a link, button, or other content element, this is another data point to consider. What's making that content element so intriguing? Is it because it's engaging, or because it's confusing and your audience needs more information about it? You can conduct further content research to find out.

Figure 10.3 is an example of a gaze plot, from Nielsen Norman Group.[1]

FIGURE 10.3
An example of a gaze plot, from the Nielsen Norman Group website.

1   Nielsen Norman Group, "Eyetracking Setup," www.nngroup.com/articles/eyetracking-setup/

Five-Second Tests

Five-second tests are short, sweet ways to find out, in a broad way, whether your content is making a positive impression or not, and whether your content is helping you and your team achieve the goals you've set for it. Five-second tests are frequently used as part of a website or landing page redesign, to help figure out if core messaging is getting across to your audience—or if it's not.

Here's the gist of five-second tests: You select the content that you want to evaluate. (It should be simple—not a dense wall of text or article-length!) Take a screen grab. You show it to your selected research participants, one at a time, for just five seconds. Then you ask a series of qualitative questions. What stood out, if anything? What details, if any, do the participants recall? What, in their opinion, is the purpose of this content? Did they absorb anything about the content's voice and tone?

You may be quite surprised by the responses.

One of the core benefits of five-second tests is they often indicate that the design and content team collectively need to go back to the drawing board to simplify, simplify, simplify. Five-second tests are powerful on their own, though if needed, you can combine five-second tests with heat maps or gaze plots or other quantitative or qualitative research to tell a fuller story about what's impressing your customers, what's confusing or not obvious to them, and what needs to be improved to help the digital experience succeed. If you're ever feeling an overload of stakeholder input, a five-second test can be your best friend in helping you gather customer feedback that ultimately helps you distill content down to a simple, single core message that you're intending to communicate.

Competitive Analysis

Competitive analysis is exactly what it sounds like: comparing your content to the competition's. However, competitive analysis has a way of taking on a life of its own. What at first glance appears to be a research project that could take a couple of days can all too easily mushroom into a weeks-long slog. Keep the scope-creep in check! The key to useful, actionable competitive analysis is threefold:

- **First**, accurately identify your competition to the best of your ability and keep the list of competitors to a minimum. Ideally, compare your content only to your Most Important Competitor's.

- **Second**, make sure to identify your analysis criteria or framework before you start the work of comparing your competition's content with yours.
- **Third**, rein in the scope of the comparison work itself, so that it doesn't stretch out for weeks and become overwhelming.

Identify the Competition

First, who is your main competition? How do you know? Do you have one main competitor, or three—or, perhaps, one or more main competitors for each product? It's worth checking in with your marketing department and stakeholders from other teams (UX research, product management, software development, product design, and especially business development or business planning) and prodding each of them a bit. Ask them to provide details about who they think the main competitor is, how and why that competitor was originally identified, and by whom. (You may be surprised that it may have been identified by an outside marketing agency that may not necessarily be as familiar with your business or your customers as the product team is!)

And just because a company was your main competitor last year, that doesn't mean they are still your prime competitor now. Markets change, products and features change, and customer preferences evolve. You need to stay on your toes and not assume that what held true in the past still holds true today. It's growth mindset for the win!

Keep this list of competitors short and scoped. Your company certainly may have a dozen competitors. But you need to pick only one or two of them for your analysis, so you can complete the work in a reasonable amount of time and, more importantly, identify insights that can help you update and improve your content. (Keep the list of competitors who don't make the cut—you can evaluate their content in future months or quarters.)

Identify the Content-Quality Criteria

If you took the time to lay the foundation of identifying your Most Important Content and your content-quality criteria (as outlined in Chapter 3, "Identify Your Content Quality Principles"), that work will be put to good use here.

Determine what you need to compare, or the altitude at which you'll be conducting your evaluation. There are three basic levels or altitudes: macro analysis, content-type analysis, and micro-level analysis.

Macro-Level Competitive Analysis

Are you most concerned about the overall landscape of your content and how it's perceived holistically against your competitors? This is a *macro-level analysis.* (The macro-level approach is typically taken by team leaders who are responsible or accountable for departmental-level content metrics and reporting them to the C-suite.)

This type of analysis may be done on a regular basis—perhaps quarterly or annually. You can compare the initial or baseline measurement to subsequent measurements. (This is one of many ways you can show the impact of your content efforts, such as when you're seeking additional staffing or budget support for your content team.)

How do you do macro-level analysis? First, take a representative sample of your company's content and a similar sample of your competitor's—maybe three or five content pieces or assets, like the home page or your app starting screen, your "About Us" page, and a key landing page that's essential to your business. (By "essential," it means that you and your leadership team are closely tracking its customer engagement and content analytics.) Then systematically compare the two sets. You yourself can conduct this sort of analysis without getting research participants involved. Or, if you have time, you can involve research participants from your target audience, to help reduce bias. Ten or 20 participants is plenty, by the way.

If you're involving research participants, it's tempting to simultaneously show them both sets of content—yours and the competitors'. But this approach can be asking for too much from them, especially if the content pieces include a lot of text. It also introduces bias.

A more successful approach is to present the content separately: show your company's content to one group of research participants. Then, separately, present your competitor's content to a different participant group. By using two separate groups, you're also less likely to see the participants' comments and insights skewed when they invariably start comparing and contrasting the two sets of content.

For both groups, ask identical questions. Some questions to consider:

- What, if anything, do you like about this content?
- What, if anything, do you dislike?
- Does anything specific capture your attention first?
- Is anything unclear? Is anything missing?
- What would you suggest be changed, if anything?
- You can add some additional questions using the research techniques outlined in Chapter 7, "Craft Your Content Research Questions." For example, "On a scale of 1 to 9, with 1 being not at all helpful and 9 being very helpful, how helpful would you say this content is to you?"

At this point, you can evaluate, compare, and contrast the responses. You may need to create a coding matrix to track the qualitative feedback. (See Chapter 8, "Analyze Content Research Results.") You'll certainly uncover gaps in your content and perhaps excesses in your content. You're also bound to get ideas for specific things to improve. (Content work is never done! A competitive analysis certainly highlights this truth.)

Micro-Level Competitive Analysis

Are you most concerned about the micro-level content, meaning for a single, specific content asset or piece? Perhaps you're launching a key campaign for a new product, and you want to know how your content might stand up against your competitor's. In the micro-level approach, you take a single content asset and evaluate it deeply. You can use the same questions listed previously for macro-level content. Add as many additional questions as you need to understand how well the content is resonating with participants (or not).

Competitive Analysis by Content Format

Maybe you want or need to know how a certain content type is faring. Do you have a podcast, a collection of social media posts, or maybe a set of related webinars? It can be very useful to laser focus on a specific content format and determine the quality of the content your company is making, compared to your competition. (This type of comparison can sometimes be a diplomatic way to make the business case for boosting your content-creation budget and staff to add variety to your content repertoire!) See Table 10.1.

WHAT IF YOUR COMPETITION ISN'T MAKING THE SAME TYPE OF CONTENT AS YOU?

What can be very eye-opening is when your competition doesn't have any examples of a specific content format, but your company has a *lot* of that type of content. For example, are you devoting a significant amount of content creation time and budget to videos, while your competitors only have a few, or don't have any? This type of content research discovery is fascinating. You need to then interpret what you've discovered, which may involve some inferences. Maybe the competition isn't as successful as your company, and they can't afford to make videos just yet. Or, perhaps, they've conducted content research that indicated that videos simply weren't appealing to their audience. (Of course, you can't know that; this is where the inferences come in.) It could be that their other content formats, such as landing pages, or their app, or social media promotions, are performing strongly for them, and they don't feel that videos are necessary at this time.

You won't always know for sure what's going on with your competitors. But you can turn inward and contemplate what disparities like this indicate about your own company and why you and your content team create the types of content that you do. Competitive analysis can highlight some awkward or even painful truths. Is there an executive who happens to have an affinity for webinars and asked the content team to make a series of them? Does your content team's skill set or comfort level around creating certain content types lead to a lopsided number of a specific content type (say, blog posts or podcasts)? It's highly likely that your publishing process or "tech stack" (meaning your content management system and other content tools) support the creation of some types of content much better than others. All these factors ultimately affect your customer experience.

By conducting a competitive analysis, you're pulling the curtain back on your own team and company's content creation behavior patterns, for the betterment of the customer experience. Is it time for a new content management system or evaluation or overhaul of your content publishing process? Use content research to gently (or not so gently!) stir things up in your company and start to turn the wheels to improve your content processes and platforms for the sake of the customer experience and strengthening of the bottom line.

TABLE 10.1   A SAMPLE QUICK COMPETITIVE ANALYSIS,
BY CONTENT TYPE

	Our Content	Competitor A's Content
Website	Wordy Complicated navigation Outdated visuals 5 call-to-action buttons (pointing to 5 product landing pages)	Concise and focused Features 1 video 2 calls to action
Blog Posts	30 5 blog posts on the same topic Some posts are more than 3 years old and include now-inaccurate information	15 Each blog post covers a unique topic Published monthly
Webinars	12 Average length: 20 minutes	None
Videos	10 videos Average length: 3 minutes	None
Help Content/ FAQs	Landing page with search field and chatbot	Complicated landing page (multiple sections with unclear organization) No chat functionality No search functionality

Meta-analysis requires a body of research that you compile over several months, or even years. There is value in gathering this body of research together and evaluating and summarizing it holistically, with a bird's-eye perspective, taking the insights from individual content research reports, looking for patterns, and synthesizing the insights into a larger, comprehensive report.

Meta-analysis is especially helpful to undertake when there are multiple people undertaking content research. A useful way of thinking about meta-analyses is that each research study can be thought of like a chapter in a book. When you collect all the chapters together, they tell a story that you wouldn't otherwise experience.

Say your small, scrappy content team conducted 10 content research studies in a year—that's a decent number. Say four of them focused on your smallest customers, or entrepreneurs. Six studies focused on owners of medium-sized businesses, which your company defines as those with up to 50 employees.

Chances are that the people who ran the studies that focused on entrepreneurs probably didn't work with or interact much with the people who ran the studies on the medium-sized business owners. Business silos affect research results. That said, when you put the research reports side-by-side, what can you and this other team learn from each other? Meta-analysis connects the dots between disparate research studies (and, often, business functions) to draw additional conclusions that otherwise wouldn't materialize.

For example, you can evaluate the highlights and verbatims from the qualitative data over time—perhaps, a year's worth. This is also known as *longitudinal analysis*. Are business owners who were interviewed or surveyed a year ago using different language to describe your product than those interviewed more recently? Overall, is there a movement toward a more positive sentiment now, compared to earlier? Is it headed into more negative territory? Or staying the same? Whichever direction you're seeing, that's a valuable business insight that will be helpful to your entire product team. If it's staying the same, that, too, is an insight worth recognizing and one that can help you and your team create your content more confidently!

You can use the same analysis technique outlined in Chapter 8. Create a spreadsheet or chart and add one row for each research report. For each report, read through and summarize: What are the standout highlights? What are a few words or phrases from the most vivid verbatims or customer quotes included in the report? This work doesn't need to be laborious or all that time-consuming. You may need only a few columns in your chart or spreadsheet to capture the most essential information from each study.

If you don't want to use a spreadsheet or chart, you can also conduct this sort of meta-analysis collaboratively, in a workshop with your colleagues, using shared documents. Make sure that each person has access to all the reports you'll be evaluating. (Collaboration apps like Miro, Mural, or FigJam are especially useful for this type of work.)

Assign a report or a handful of reports to each person and ask them to copy and paste the phrases and customer verbatims that jump out the most from each report. You can also do this summarizing by using printouts and sticky notes, if you want to use that approach. If you're using sticky notes, make sure to use a unique color for each report or notate the name of the research report at the top of each note, so you can keep track of which comment applies to which report.

Then take a step back and do some affinity mapping. What new patterns are emerging? Is there anything that's surprising to you, or that would be surprising or valuable to your stakeholders? Are you learning something you didn't realize before about your audience or customers that's only now coming into focus?

Whether or not you have startling epiphanies while conducting the meta-analysis, there is value in collecting a body of research into one handy, summarized report. Share this with your brand team, marketing, product leaders, and new hires. And use it as inspiration for future content research.

Some kinds of "golden nuggets" that you may uncover from meta-analysis: audiences that you *thought* were tech-savvy and sophisticated may *not* be. Audiences whom you previously thought understood specific product descriptions or language may require more detailed, nuanced definitions for greater clarity. Features that were created specifically for one version of your product—such as enterprise software—may be very valuable for users of the "small business" version of your product.

SUS

System Usability Scale or SUS is a quantitative measurement that's a tried-and-true user experience measurement framework, and therefore worth including in your content research tool kit. To measure System Usability Scale (SUS), you present a specific piece of content or scoped content experience to a sample audience; then ask them a standardized list of 10 scale questions—using a rating or Likert scale of 1 to 5—that assesses various elements of the UX, and how easy or challenging the user feels they are. (SUS was briefly addressed earlier in the book in Chapter 7.)

SUS is a helpful framework to take a baseline measure of a key user experience—your Most Important Content—and then reassess it on a regular basis, such as quarterly or biannually, to understand how well it's landing with your customers. UserTesting includes SUS testing and scoring capabilities for some versions of its product; SUS is also available for licensing from MeasuringU (www.measuringu.com).

A couple of things to note about SUS. First, it is *not* a purely content-focused measurement. Most of the questions are general, addressing the overall user experience. SUS primarily evaluates the combined experience of content *in tandem with* design. Only one of the SUS questions focuses on content by itself, with the intent of evaluating content clarity. In addition, the scoring of SUS is complex. Scores are based on a range of 0 to 100, but are *percentile rankings* (not percentages), which can be tricky to comprehend. *An above-average score on SUS could be as low as 68.* Be careful to clarify the quirks of the scoring model when presenting SUS scores to stakeholders and executives, or they may understandably feel angst about the relatively low numbers. ("Only 68? If that were a grade, it would be a D!")

While SUS is obviously not the simplest of evaluation methods, there are many benefits of using it for your content team:

- It can convey whether the user experience being evaluated is above or below average, from an industry perspective.
- If you take a baseline and then track a specific UX measurement over time, it can show you the degree to which the customer experience is improving (or not). It can also flag where investments in time, budget, and staffing are necessary.

- Its "usability" and "learnability" measurements, which evaluate how easy the respondents feel the experience is to use on their own (without requesting help from your tech support team, for example) are fascinating measurements that you can use to improve your customer experience, especially when it comes to content clarity. SUS can often flag content gaps—where details or entire topics are missing, and necessary to support customers in their Jobs to Be Done.
- SUS can be used on very small sample sizes. MeasuringU recommends conducting SUS research with as few as five users.[2]

Figure 10.4 shows a sample output of an SUS evaluation.[3]

Sample System Usability Scale (SUS) Output

RESULTS

Question 1	Question 2	Question 3	Question 4	Question 5	Question 6	Question 7	Question 8	Question 9	Question 10	Question 11	Question 12	SUS	Usability	Learnability
2	1	4	1	3	1	5	1	5	2	Best imaginable	10	82.5	81.3	87.5
3	2	3	1	3	2	4	3	4	1	Good	8	70.0	62.5	100.0
5	1	5	1	5	1	5	1	4	1	Excellent	10	97.5	96.9	100.0
4	2	4	1	4	2	4	2	4	1	Excellent	7	80.0	75.0	100.0
5	1	5	1	5	1	3	1	5	5	Excellent	10	90.0	100.0	50.0
4	3	4	1	2	1	4	2	5	1	Excellent	8	77.5	71.9	100.0

FIGURE 10.4

Here is a sample output from an SUS evaluation with six participants. Most respondents found this experience to be excellent; the second row shows a participant who found the usability to be difficult (note the overall usability score of 62.5).

---

2   Jeff Sauro, "10 Things to Know About the System Usability Scale (SUS)." June 18, 2013, https://measuringu.com/10-things-sus/.

3   Trevon Gripper, "SUSplus App," May 12, 2022, https://help.usertesting.com/hc/en-us/articles/360053994052-SUSplus-APP#

Another Research Option: The SUPR-Q

If the scoring of SUS is too complex or confusing for your stakeholders, SUPR-Q is another option. SUPR-Q stands for *Standardized User Experience Percentile Rank Questionnaire*. SUPR-Q is similar to SUS, but it has a shorter list of standard questions: 8 questions for SUPR-Q, compared to SUS's 10. SUPR-Q scores slightly different elements: usability, appearance, loyalty, and trust/credibility.

SUPR-Q's scoring is simpler and easier to understand, with 50 being the 50th percentile. SUPR-Q provides a helpful composite score that tells a quick story for your team and stakeholders, and, like SUS, can be easily tracked over months or years to gauge ongoing improvement (or dips in UX quality) over time.

What SUPR-Q and SUS can't tell you, however, is what elements or words in your content experience are working for your audience, and what aren't. You'll need to do additional content research using heat maps, or other quantitative and qualitative analysis, to find out. The SUPR-Q tool is also available from MeasuringU.[4]

Social Media Research

Social media platforms can also be leveraged to discover useful content insights. (The ethics surrounding these various platforms is not a topic that's going to be explored here.) If an online testing platform is out of your budget, or you're in need of very quick research results, you can take advantage of the built-in survey tools on LinkedIn, Facebook, or Instagram to gain insights about your audience. The targeting may not be as refined as it would be with an online testing platform, but the survey results and comments can still provide quite helpful input as you create and refine content.

One word of warning with these tools: As with all content research, it may be a remote risk, but it's a risk nonetheless to request feedback on content if it's focused on a product or feature that's a work-in-progress and has not yet been made public. It's easy for participants to take a screenshot of the content that you show as part of the research and then share it with their followers. You can reduce the risk somewhat if you use a social media platform like LinkedIn or Facebook, which have settings so that you can make your company's feed open only

---

4   Jeff Sauro, "10 Things to Know About the SUPR-Q," June 13, 2018,
    www.measuringu.com/10-things-SUPRQ

to its followers. But to be on the safe side, it's best to use social media research for refining and iterating content that's already customer-facing, and for products and features that have already launched.

Instagram Stories

Instagram Stories are collections of videos and photos that are intentionally short-lived, remaining online for only 24 hours. With the permission of or help from your company's social media team, you can use Instagram Stories to gather quantitative and qualitative feedback in a relatively easy and quick way from people who follow your company on Instagram (see Figures 10.5 and 10.6).

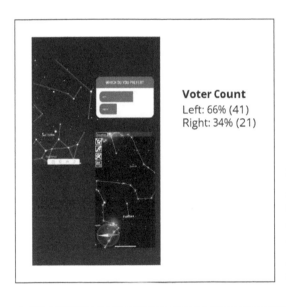

Voter Count
Left: 66% (41)
Right: 34% (21)

FIGURE 10.5
A sample Instagram Stories survey question asks respondents to select which of two app layout prototypes they prefer.

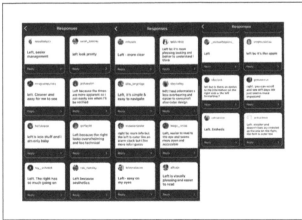

FIGURE 10.6
The qualitative (or "why") comments from Instagram Stories respondents provide feedback on why they preferred one version of the app over the other.

Moderated Interviews

Moderated interviews are another terrific tool for your content research tool kit. They gather primarily qualitative information, so are therefore labor intensive and time-consuming to conduct. However, the effort pays off greatly because you're gathering input and feedback that's straight from your customers' mouths and can have dramatic influence and impact on your content.

"Moderated" simply means that someone (quite possibly you!) is present as a research participant answers questions. The beauty of moderated interviews is that, unlike with online research platforms, you're able to edit or add questions on the fly in response to the answers and information you're hearing from the participant.

Preparing for Moderated Interviews

The first step in running moderated interviews is to use your typical content research study template and scope your research as best you can, determining your target audience/customer.

Next, write the short list of questions that you're *planning* on asking participants. Depending on how complex the user experience is that you're evaluating, it's generally wise to keep this list to a short, scoped list of five questions or fewer. (Similar to online surveys or online research platforms, your participants' attention span will wane over time, and aiming for a maximum interview length of around 20 minutes will help keep their focus sharp and therefore their responses as useful as possible.) Work with your small group of core stakeholders to develop these questions, so everyone's on the same page.

Like with writing multiple-choice and scale questions, be careful not to introduce bias into the questions themselves. (See Chapter 7.) Be sure to also follow the "Company Policies, Legal Requirements, and NDAs" section of Chapter 6, "Research Planning and Stakeholder Management," about working with your customer experience and legal teams to ensure that your research plans align with company policies as you recruit participants.

If your participants are amenable to it, you may want to conduct moderated interviews with your subject-matter experts, like the product manager, your product team colleagues, or the lead product designer participating with you. You can also offer to have these colleagues attend the interview session, but as listeners only.

Conducting the Moderated Interview

You can conduct moderated interviews in person or over online videoconferencing platforms like Zoom or Microsoft Teams. The online research platforms like Lookback and dscout also enable moderated interviews. When running moderated interviews, before getting started, be sure to obtain written permission from your participants when recording any audio or video. (In 11 states, it's illegal to record conversations without both parties' consent—California, Delaware, Florida, Illinois, Maryland, Massachusetts, Montana, Nevada, New Hampshire, Pennsylvania, and Washington.) If your legal team requires it, be sure to have participants sign a Nondisclosure Agreement (NDA).

Some participants may become wary of the formalities of these forms and legal permissions. Do your best to help them understand that they are required by your company, and that the interview itself should feel relaxed and conversational.

Before you get started, clarify with your participant how long you expect the interview to run (20 minutes or so is ideal). Also provide your contact information (phone and email) for them to ask any follow-up questions or provide additional comments and feedback if it occurs to them after your session is over. Creating connection with the interviewee is important to "warm them up" so their feedback and ideas flow during your short time with them.

As you ask questions during the interview, stay on your toes and prepare to probe a bit if you hear a response that is unexpected, intriguing, or otherwise piques your interest. Gently ask the participant to provide an example or additional details, or otherwise provide color and context to their responses.

Keep your eyes on the clock (without it being too obvious to your interviewee). You may find as you go "off script" and add additional follow-up questions that the time flies, and you may not have enough time to get to all the original questions on your "scripted" list.

For further details on interview techniques, see "How to Moderate User Interviews" from the Interaction Design Foundation.[5]

---

5   Ditte Hvas Mortensen, "How to Moderate User Interviews," Interaction Design Foundation, 2000, www.interaction-design.org/literature/article/how-to-moderate-user-interviews

Analyzing Moderated Interview Responses

The process of analyzing moderated interviews is similar to the qualitative research analysis steps outlined in Chapter 8. If you use an online research platform and it provides a transcription of the interview, that's a huge time-saver. Transcribing a full interview (or series of interviews) by yourself is often too time-consuming; what you can do instead is make sure that you accurately capture the most illuminating comments and feedback, especially for when you went "off script" and dug in with deeper questioning.

The "verbatims" or direct quotes from moderated interviews will provide much food for thought for your whole product team, especially product managers and product designers. Sometimes the feedback heard in moderated interviews can prompt product teams to significantly alter product or feature plans or even scrap them altogether, go back to the drawing board, and start over with plans that reflect the "voice of your customer" as heard in these interviews.

Making It All Work for You

Heat maps and gaze plots require special software, although it's often low cost. Heat maps and gaze plots have a high return on investment because they identify where customers' attention is focused within a specific screen, or not focused at all—which helps you uncover usability insights that would otherwise not be obvious.

Competitive analysis is useful because it can be done quickly and easily by a single person working alone. SUS and SUPR-Q are powerful, sophisticated, thorough research frameworks that require more investment in time and effort—and may require partnering with your colleagues on your UX research team—but are powerful tools to use to benchmark and show incremental improvement in your content experience over time. Social media research is quite the opposite of SUS and SUPR-Q; it can be performed quickly and relatively easily. With this wide variety of tools at your disposal, you can understand your users' needs and create effective, resonant content that serves their needs well and supports your company's business performance.

Apply Insights and Share Business Results

Here's where content research shows its power: taking the insights and knowledge you discovered through research and then updating the relevant content so that it reflects these newfound discoveries. Just like with every other step in the content research process, there's a smooth, helpful, collaborative way of going about this work and where you do the due diligence to show the impact of your content research efforts. On the other hand, there's the other road with bumps and potholes that are best avoided, and where you neglect to show the business results related to your work (or unfortunately let others take credit for your hard work). The following tips will help you take the easy road *and* get recognized for your work!

Work Within Your Product-Development Process

As long as you have your final quantitative and qualitative content research results, you can start taking the needed steps to update the content in your customer experience. Just as you don't want to share your research results when they're incomplete, you certainly don't want to start updating your app or website to reflect incomplete research results.

Balance Urgency with Process

Do your best to implement your research discoveries as soon as possible—without overly disrupting your team's product development or content publishing workflow, that is. It's exciting to have fascinating research insights at your fingertips that you're confident will have a positive influence on the customer experience and your business. That's no reason, however, to act like a special snowflake and upset your product team's regularly scheduled product sprint in order to update content to reflect your research insights. Align with your team's typical work processes as much as possible!

If you yourself have access to your content management system and have the necessary permissions and skills to update and publish customer-facing content, then go ahead and do so. Just make sure that any people who need to know about the updates—and the timing of the updates—are informed. This group of people includes the core team of stakeholders who were involved in your research planning. It also includes a broader group of colleagues, including your data analysis team, marketing and marketing operations, and any product managers beyond those who were part of the planning process—as well as your localization/translation team, if you have one.

If, on the other hand, your team does more gatekeeping and has more structured engineering or publishing processes and workflows, you'll need to work within them and coordinate with the right colleagues to make your content updates go live. If you work with Agile workflows and have regular product-development sprints with work clearly identified and prioritized prior to each sprint's start, you may need to have a candid conversation with your product lead about how best to get the content updates published. (A quick bit of advice: If you're conducting content research on an ongoing, regular basis—which is ideal—you'll want to budget time and capacity into each sprint or each month's planned work to account for the time and effort needed for the research prep, the research work itself, and any necessary content updating.)

Whichever way you work out the details of your post-research content-update process, make sure to be mindful of the feelings of your teammates. If your research is iterating content that's already been published and is customer-facing, be sensitive to the feelings of the person or people who created the original content. Research can be considered a form of feedback, and critical feedback can be tough to absorb. And be mindful of the feelings and workload of all your product teammates, as sometimes the excitement that content research generates can overshadow other work in progress. Everyone's working hard, so be kind and thoughtful and thankful to your teammates.

Note the Date When You Make Content Updates

If your content management system or publishing tools don't automatically notate the date when content updates and changes are made, then you need to make a note of these dates manually. Keep track of the day and time when your research-informed content updates went live and were made customer-facing. That way, you can clearly measure the business impact of your content changes (see the following section, "Show the Impact of Your Updates"). As part of your content research practice, you can start a spreadsheet of content research plans, reports, summarized insights, go-live dates for the iterated content, as well as notes that document the business impact.

WHEN TO UPDATE WITH URGENCY

Back in Chapter 1, "The Power of Content Research," I shared the story about working for a health insurance company that sold three different types of insurance plans: Gold, Silver, and Bronze. Barely any customers were buying Silver plans. The digital experience team learned, through content research, that customers were utterly misunderstanding what the Silver plans offered, mistaking them for Medicare plans meant only for people over the age of 65. Customers were ignoring the Silver plans in favor of the Bronze or Gold plans—and quite likely under- or over-insuring themselves and their families in the process.

In the case of this content research insight about Silver plans, it did make sense to act with urgency and make content updates as quickly as possible. We brought all hands on deck to immediately update the home page content to explain that Silver plans *had nothing to do with Medicare*. This was a business emergency with a potentially devastating impact for the company, as these plans were sold only for a 10-week period each year to abide by government guidelines. And they are a major source of company revenue.

To address the customer confusion and content insights that we discovered about Silver plans, we called an emergency meeting and quickly coordinated with the right people across the company (and many outside vendor teams) to update the website, app, marketing communications, social media posts, and customer-service scripts. We acted with urgency and dropped all other work that day to make the changes happen and communicate internally throughout the company about this change. These changes were made easier because we shared direct quotes and feedback from customers that were obtained during the content research process.

It's not every day that content research helps you discover something as surprising as this insight about Silver plans. But you too may discover similar confusion on behalf of your customers and need to make content changes with urgency. Make sure to set the bar high about such urgency, so as not to wear out your welcome with product teammates. And be sure to be kind and show appreciation toward your colleagues as such updates are made.

Show the Impact of Your Updates

After your new, improved content is published, you can use content analytics to show how well the new, research-informed content is outperforming the previous version.

Say you work for a company called *Dogwa*, which sells vitamin- and mineral-fortified water for dogs. Your home page and social media campaigns were performing OK, but were not spectacular. You therefore ran a research study to determine which specific adjectives your target audience felt were most appealing to use when describing Dogwa. Then you took those adjectives and updated your home page and social media campaign content accordingly.

Calculating Improvements in Customer Engagement

You can then measure the difference (or "delta") between content performance before research and content performance after research. To do this, take the content performance data that shows how well (or not) your home page, social media posts, and other content formats were engaging customers prior to research. Then update the content with the new adjectives identified during research. Wait a bit and then re-measure how well the refreshed content is engaging customers.

This difference or delta will not be able to be measured immediately. You may need to let the new content "cook" for a little while before being able to measure any change. The time you will need to wait depends on how many visitors you get, on average, visiting your home page or app or viewing social media posts on a daily or weekly basis. (A day or two is all it takes for websites with tens of thousands of daily visitors; depending on the volume of customers who view your content, you may need to wait a week or a month to clearly illustrate the delta.) It's important to take the time to "bring the research home" and illustrate the return on investment (ROI) or value of the time and effort that was spent running and analyzing research and implementing the insights.

Table 11.1 shows a hypothetical content data chart that illustrates the positive difference that content research created in just one month's time for Dogwa, rendering both the home page content and social media content more engaging to customers.

TABLE 11.1 DOGWA CONTENT RESEARCH: IMPACT ON CUSTOMER ENGAGEMENT

Date	Home Page: Average Daily Click-Through Rate	Social Media Posts: Average Daily Click-Through Rate
Jan 1, 2023 *Before content research*	10.5%	4.3%
Feb 1, 2023 *After content research*	15.2%	7.2%
Difference	+4.7%	+2.9%

These numbers are impressive by themselves. But they'll be more impressive if you can tie them to results that are even more concrete (especially if you can connect them to results that pique the interest of senior leadership and your C-suite). Depending on how detailed your content data is, you may be able to extrapolate further from these customer engagement metrics to dollar bills—those all-important revenue metrics.

If you know how many customers visit your site each day, and what percent of home page visitors makes a purchase, as well as the average amount spent by each customer who makes a purchase, then you can calculate the net increase in revenue associated with your content research. Table 11.2 illustrates just such a calculation.

At first glance, a daily increase of $9,400 may not seem like a huge lift. But in a 30-day month, that's $282,000. In a year, it's an annual increase in revenue of $3,431,000. That's a terrific return on investment and testament to how words matter greatly when it comes to user experience.

Take Credit for Your Business Results!

How you go about sharing the results of your content research will depend on how your organization measures and tracks business impact. If your organization holds a monthly business review meeting, by all means, raise your hand to volunteer and share your impact storytelling in that forum.

TABLE 11.2 CALCULATING REVENUE IMPACT FROM CONTENT RESEARCH

	Daily Home Page Visitors	Daily Av. Click-Through (Engagement) Rate	Total # of Engaged Customers	Average Purchase Size	Daily Average Revenue
Before Content Research	100,000	10.5%	2,100	$100	2,100 X $100 = $21,000
After Content Research	100,000	15.2%	3,040	$100	3040 x $100 = $30,400
Difference or Delta		+4.7%	+940 or 940/2100 (the number prior to research) Multiplied by 100 = +44.7%		30,400 – 21,000 = $9,400

You may find that content designers and content strategists aren't typically invited to such business meetings. If that's the case, it's time to shake things up. You led this content research, you implemented the insights, you measured the business impact, and you should get credit for it. Your product manager may offer to present the results for you, as PMs often attend these business review meetings. Don't acquiesce to having someone else represent your hard work. As mentioned earlier in the book, content research is a way to build more respect for the practices of content design and content research. Sharing the business impact of your research is a prime way to build respect for your team and expand its influence. (If you feel out of your comfort zone presenting at a meeting like this, be brave! Be proud of the results you drove, and frame your presentation around how you've helped your company's digital experience become more customer-centric and effective.)

Update Your Content Guidelines/ Style Guide

In addition to sharing the business results of your content research and content updates, make sure that the knowledge you've gleaned from research is available to others in your company for future reference. Update your style guide or word lists (and if you're in content design, your content design library components) to reflect your content research insights. For example, if you learned that "vital" was the adjective that most intrigued your Dogwa customers, add it to your style guide. Include an annotation of some sort—an icon or mark perhaps—to denote that the word or phrase was identified through content research.

You can get very detailed with your annotations, mark the date on which the research was conducted, and indicate when the research report was completed. Depending on the platform that's used to build your style guide or design library, you may be able to include links in your style guide or library that point to your content research report *and* to the test itself within your research platform (such as dscout or UserZoom).

The results of annotating your style guide or design library are manyfold. First, any and everyone who refers to these frequently used resources will be made aware that content research is something that happens at your company—and that it happens on a regular, ongoing basis. Second, it's also a great way to help inform new hires about your content research program. Third, it serves as an inspiration and an ongoing reminder to your content team that it's important to validate as much of your Most Important Content as possible. (You may count your style guide and design library among your Most Important Content.) On top of all these wonderful things, promoting your content research efforts builds trust and respect for your content team. Figure 11.1a shows an example and Figure 11.1b. shows a closeup close-up of a writing style guide's word list where specific words and phrases are flagged as having been validated through content testing. ("Let's Be Clear" was the company's plain-language initiative in which jargon was replaced by simpler terms whenever feasible.)

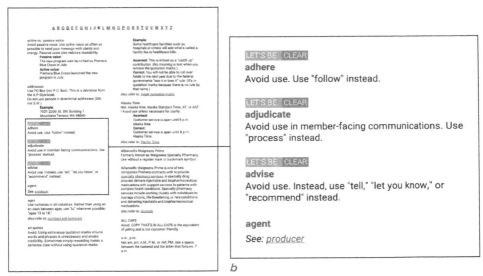

a

b

FIGURES 11.1a and b

An example of a content writing style guide that highlights the words and phrases that were evaluated using content research. In this case, words or phrases that have been evaluated with content research are marked with the "Let's Be Clear" icon.

PRO TIP INCLUDE THE WORDS *TO AVOID* IN YOUR STYLE GUIDE

When you update your writing style guide or content design library with information and details from content research, be sure to include the words or phrases that your research indicates ought to be *avoided*. Which words were confusing to your audience, or weren't engaging for other specific reasons? Which words or phrases maybe even offended your audience (as Tracey Van-tyghem found in her case study about eradicating jargon that's included in Chapter 5, "Identify Your Content Research Goals"). When it can be helpful, include a short. pithy quote from the qualitative research questions to emphasize the importance of avoiding these words, along with a link to your research report.

Inform A/B and Multivariate Experiments

Your content research practice is also a boon for any A/B or multivariate experimentation that's happening in your company. Remember, as mentioned in Chapter 1, it's often preferable to conduct content research instead of "live" content experiments like A/B or multivariate experimentation. *This is because content research is much less risky, less resource-intensive, and often able to provide insights more quickly than A/B or multivariate experimentation!*

If your organization is quite eager to conduct A/B or multivariate experiments, you can advocate for a round of content research prior to starting any experimentation. In other words, you can use words and terms that "bubbled up" during content research in A/B experiments as "experimental" versions of content to be tested against the control version. That way, you're "hitting the ground running" with a pre-vetted version of content as your experimental version.

Running content research ahead of A/B experiments is a best practice that's worth emphasizing to your product team and other stakeholders. A/B experimentation is very popular among product leadership because it clearly measures improvements to the customer experience and your business's bottom line.

As the content research lead, it's your role to help your peers understand that while A/B experiments are valuable, they may not be necessary. This is a radical concept for many product teams. Content research can often uncover what words work best for your customers, *without the risk* of experimenting on live, customer-facing experiences. If you conduct content research instead of A/B experiments, you can still calculate the impact that the research had on the customer experience and your company business results and revenue.

To do so, you merely need to make a note of the exact date and time when you updated your customer-facing content with the content that was identified through research as the content that's most engaging to your audience (or preferred or clearest, or whatever superlative, depending on the type of research you ran). Give the customer-facing content a few days, weeks, or a month or so after it's published (shorter periods of time for websites and apps with lots of daily visitors; longer if your site is less heavily trafficked). Then you can use your content analytics to show the difference in performance between the content before research and the content after research.

This approach may not be as exciting as live A/B experiments, but it's less of a risk to your business and brand. Another point worth calling out is that A/B experiments can also create customer confusion. Have you ever been "bucketed" into an A/B experiment and confused by seeing one version of content at one point in time and then a different version of content the next hour or next day? A/B experiments can sometimes reveal your company's "dirty laundry" when they're not seamless to your audience. A/B experiments carry a cognitive load to them that's often brushed aside by the product team that's running them. You can lean on "noninvasive" content research instead to identify which words work most effectively for your audience.

Making It All Work for You

Updating your customer-facing content using the insights you uncovered through research feels super satisfying! However, for the sake of your content team's reputation, and for the success of future research studies, it's important to make sure that you arrange to iterate or update your content in a way that doesn't disrupt the day-to-day functioning of your entire product team. Make friends with your product manager, engineers, or content producers whom you will need to work with to push those content updates live.

While you're bringing your content insights to life, get organized and make plans to measure the impact of your research efforts, extrapolating impact into revenue and clear, concrete business performance numbers whenever possible. Also be sure to communicate those results with the people who want and need to know about them, in particular, your senior leadership team. As you share these stories of business impact that's created directly from content research, be prepared to get requests from around your company for you and your team to run more and more frequent content research studies.

CHAPTER 12

Build Momentum with Content Research

W hen you run content research, you'll be kept on your toes. Your hypotheses and predictions often get blown out of the water, so you'll find that you and your team will continually need to check your biases—all very good things.

At the same time, as you run research studies, uncover insights, and share them across your company, people in and outside your immediate team will take notice. They'll start asking when the next content research project is kicking off, how they can get involved, and what they can do to learn how to run content research themselves.

All this positive energy and enthusiasm feels terrific! But there's no denying that content research requires time, energy, and know-how so that it's done right—supporting your content team goals and standards, and your business's goals. The following tips are intended to help you with the balancing act of building and sustaining energy for your content research program, while creating boundaries to keep your content research practice from overwhelming the rest of your day-to-day work.

Document Your Research Methods and Tools

If you haven't done so yet, a key first step to building momentum for your research practice is to create a one-stop shop for your content research tools. This site will help you and your current coworkers and provide future new hires with the information they need. Create a tool kit on a wiki or other shared site that your entire content team can access. Include your content research test templates, any internally or externally created documentation on how to use (and obtain credentials for signing into) your testing platform, a section for completed research reports, and your spreadsheet or deck for sharing ongoing business results and impact that was informed by content research. (See Chapter 9, "Share Results from Content Research Studies," for guidelines on how to standardize your file names, to make it easy for completed research plans and reports to be searched and sorted.) Make sure that as new people join your team and company they're aware of these resources and tools. (One of the most common issues with research teams of all sorts is when resources are wasted by "reinventing the wheel" by repeating research studies or running research studies that are very similar to those that have already been completed.)

Create a Content Research Roadmap

What do you want your content research program to accomplish? It could be any number of things:

- Get your team started with content research.

- Generate greater awareness of and respect for your content team and its work.

- Provide evidence of clear business impact that can be directly credited to content research.

- Make the case for additional content team staffing.

- Help your content and product team achieve their OKRs (objectives and key results) and goals for the quarter or year.

- All of these things!

Think hard about what exactly it is that you and your content team need. For example, what is it that you need to achieve the business goals you've set, what your content team members need to reach their career goals, and what *all* of you need to *make your work as content professionals easier and more fulfilling*? Then make an organized, methodical plan and timeline for achieving these goals through a regular cadence or "drumbeat" of content research. Make sure that your content research goals are achievable, yet still ambitious. This kind of roadmap is something you can refer to at weekly one-on-one meetings and your monthly business reviews, to build a sense of accountability and responsibility across your team.

Which content assets need to be clearest and easiest to understand, for the sake of your customers and the success of your company? Which ones receive lots of visits, yet haven't had their content quality carefully evaluated in recent history (or, perhaps, ever)? Make a prioritized plan to evaluate and iterate your Most Important Content and any brand-new content for key upcoming product or campaign launches. Use content research to validate that the words being used in customer-facing content reflect the language used by your customers. Check to see that the words are as clear and engaging as you can make them, and ultimately will create the strongest customer experience and business results possible.

In Figure 12.1, you'll find a sample content research roadmap for a large team as it was getting started learning how to do content research.

Depending on the volume of studies your team is running, you may not need to be concerned with your team's remaining "balance" of research studies. Or you may find you're running so many tests that you'll hit your limit well before your subscription's timeframe is up. It can help to track your content research efforts in whatever workflow tool your team uses (Asana, Jira, Smartsheet, Azure DevOps, etc.) and assign a content team lead to track the team's overall research study "burndown" rate.

You may find you'll need to slow the pace of your content research program. Or you can take the business results that are directly tied to your team's content research studies and use those improved revenue and customer engagement numbers to make a business case for renewing or upgrading your content research platform.

Remember That Content Needs Governance!

Ideally, your content team has a governance process. In other words, you don't simply publish content and call it good. You revisit your Most Important Content on a regularly scheduled basis, to make sure that it's accurate, clear, and as engaging as possible. Content also needs care and feeding to make sure it's up-to-date, aligns with your voice and tone and branding standards (which can evolve over time), and otherwise reflects the content standards that you and your team have documented. (For more details about this topic, see Chapter 3, "Identify Your Content Quality Principles.")

Make a plan for revisiting the content that you evaluated and improved using content research—quarterly, biannually, or annually, depending on your team staffing and resources, and how the content is prioritized on your list of Most Important Content. Conduct additional rounds of research to find out whether the content is still working, or whether it needs tweaking to help it communicate as clearly as possible, in language that your customers are comfortable with and understand.

Train Your Whole Team

Sustaining a content research practice takes a village. Avoid becoming the sole content researcher on your team or in your company! While running content research is energizing, if you do find yourself as the one and only person who knows how to conduct content research, you'll run the risk of becoming burned out. Not only will you feel a sense of being overwhelmed by the sheer volume of research work you as a content lead need and want to do, but you'll also find people across your company from all practices—product management, marketing, and more—will continually reach out for "just one really quick, short research study." That's an unsustainable position to be in.

It's a Catch-22: You want to say yes to anyone who shows interest in content research and wants to use its power to improve the customer experience and to bolster the role of content in your organization. Yet, it's tough to be the only point person who organizes, runs, and evaluates content research studies.

To protect yourself from becoming overwhelmed, you need to get others on your content team (ideally, your entire team) or other Friends of Content trained and comfortable with the practice of content research. You also need to consider an "onboarding" plan for bringing new hires up to speed on how to conduct research.

The "Each One, Teach One" Method

One approach for getting your team ramped up is to approach it one person at a time, intensely and thoroughly. Once each person gets some experience and feels comfortable setting up, running, and analyzing tests, then ask them to coach one person, and so on. This approach requires a deep initial investment of time and effort on your part.

As you train coworkers, they can shadow you as you plan a research study, coordinate with stakeholders, run a study, and then analyze it for golden nuggets and report its results. After they've shadowed you, you can switch roles, and observe and gently coach them as they undertake their first research study. You can answer any questions that crop up along the way, to make sure your newly minted content researcher feels increasingly confident about running research studies independently.

There's muscle memory involved in a complex, multistep process like content research, especially given the many features available in online platforms like UserZoom. You should expect it to take a couple of practice runs before your colleague is comfortable. Don't plan on them becoming an expert instantly.

Pros and Cons of "Each One, Teach One"

One major pro of the "each one, teach one" approach is that you can immediately lighten the research load on your shoulders. This approach helps your colleague become familiar with content research techniques, the ins and outs of the research platforms or tools that you have at your disposal, and any pro tips you may have to offer about sharing results across your company.

Some cons of the "each one, teach one" approach are obvious—it requires an investment of time and can be slow going. If you work on a large team, "each one, teach one" isn't typically the best approach. However, if you have a limited number of licenses for your research platform, this may be the best or only feasible approach to take as you begin your content research program. As more studies are run, more insights are shared, and research gains momentum, you may be more easily able to obtain additional licenses for your testing platform, at which point it makes sense to scale your program by adopting the "whole team training" approach, detailed next.

The Whole-Team Training Approach

If you have a large team (over 10 people or so), you can consider a whole-team training workshop, to get the entire group oriented and familiar with the methods, options, and best practices at the same time. Because there's so much information to absorb around content research, you may want to consider splitting up the training into two separate sessions on two different days, perhaps a week apart to help those people who are brand-new to content research to be better able to absorb all the information more easily.

If you're using a platform like Qualtrics, UserZoom, or dscout, you can work with your account manager to ask them or a trainer from the company to run training sessions for your team. Such "getting started" workshops are generally included in your annual software license; be sure to check before scheduling one.

You'll need to be sure to attend this training session or ensure another experienced content research pro is there. That's for several reasons:

- These platforms weren't created specifically for the purpose of *content* research and need to be adjusted a bit to suit the special needs of content research. Your rep may be accustomed to providing training that caters to UX research only, or UX product design. If your team is not UX-focused, the tactics may not be as relevant or helpful.

- Your account rep or training lead can't possibly learn all the ins and outs of your content landscape. They need a subject matter expert to help answer the inevitable questions that will arise during the workshop:

  - Content team goals
  - Content formats
  - Stakeholders
  - Content publishing or engineering processes (or any pertinent processes or quirks around your content management system implementation)

Pros and Cons of Whole-Team Training

Whole-team training is quick, and can be fun, especially if you're able to arrange a half- or whole-day training session. The pros of this method are that it can generate a lot of energy on your team; it's exciting to see examples of content research and learn how it empowers everyone. It can make it easier for your team to get rolling with a regular "drumbeat" or cadence of research.

One main drawback is that it can feel theoretical, at least until individuals start using your research platform. If the people being trained aren't bringing real-world content challenges and setting up and running a "real" content study during the workshop, it makes it harder for the research tactics and methods to stick in their head. It can be hard to make the leap from a whole-team workshop to actual content research. Another thing to keep in mind: learning differences. Some people on your team may be auditory learners, some visual learners, and others more comfortable with written instructions. Not all workshops may accommodate these various learning preferences.

To help mitigate the flood of questions you may receive following an all-team workshop, it's a good idea to have at least one other person trained and proficient at content research ahead of the all-team training. Then you can designate that person as another "go-to" person or subject-matter expert to answer the questions you'll get from your team following the workshop, as they go about the work of creating their own research study.

NOTE PUT WHOLE-TEAM TRAINING TO WORK QUICKLY

It's easy to participate in a training session, have a few weeks pass by, and then when you go to use the information you learned, it feels like you hadn't taken any training at all. For content teams, this risk is all too real, as they are often understaffed.

To help prevent this knowledge loss after a whole-team training session, ask each person on your team to do their best to complete their first content research study within two weeks following the training workshop. That way, they're more likely to have details from the training fresh in their mind, and they can immediately start building the muscle memory needed to get proficient at content research.

Include Research in Your Product Planning

To support the success of content research, you need the support and advocacy of your stakeholders and partners in product management, product design, software, and analytics. If your team holds regular workload and prioritization planning sessions (monthly, quarterly, biannually), make sure to voice the need to carve out time and resources for content research. Working content research into your organization's product development planning and regular sprints and earmarking the hours and bandwidth needed to do content research well is the ultimate way to support your content team's work in this space.

It may be difficult to achieve this at first, but once you have several success stories with data around revenue, customer engagement, and other business success metrics, you'll have a captive audience that understands the power of content research to help your org achieve its business goals. As the workflow tool in Figure 12.2 shows, it doesn't take a lot of time earmarked for content research to make a tremendous difference in your content performance and by extrapolation, your company's bottom line.

Sample Content Team Sprint Capacity

	NAME	PROJECT	STATUS	START DATE	END DATE	DAYS ALLOCATED
Senior content design lead	Jackie T.	Content research	In Progress	07/01	07/31	3
Senior content design lead	Jackie T.	Feature ABC launch	In Progress	07/01	07/31	6
Senior content design lead	Jackie T.	Feature XYZ development	Complete	07/01	07/31	6
Senior content design lead	Jackie T.	Feature 123 iterations	Needs Review	07/01	07/31	5
Senior content design lead	Jackie T.	Monthly analytics work	Overdue	07/01	07/31	2
Content designer	Kay J.	Feature 123 launch	In Progress	07/01	07/31	6
Content designer	Kay J.	Feature 456 launch	In Progress	07/01	07/31	6
Content designer	Kay J.	Content research	Complete	07/01	07/31	3
Content designer	Kay J.	Fit & finish, Feature 789	Complete	07/01	07/31	5
Content designer	Kay J.	Monthly analytics work	Not Started	07/01	07/31	2

FIGURE 12.2

This product content team sprint spreadsheet shows how each content team member allocates some time each month to content research.

Participate in Meetups and Conferences

Communities of Practice exist all over—they don't necessarily need to be confined to your company only. Check out the meetups in your area (www.meetup.com) that are focused on the topics of content design, content strategy, and usability. They are always looking for folks who are brave enough to share their own experience for the benefit of others and for the overall practices of content design and content strategy.

Conferences like the Design & Content Conference (https://Content .Design), UXDX (www.UXDX.com, which holds events internationally), Content Confab (www.ConfabEvents.com), and Button (www.ButtonConf.com), as seen in Figure 12.3, are other great places to share your experiences with content research. Confab and Button were founded and are run by Kristina Halvorson, the co-author of *Content Strategy for the Web* and the founder of the Minneapolis content strategy agency Brain Traffic.

FIGURE 12.3

The Button Conference home page (**www.ButtonConf.com**) touts that you can "come find your people." It's true that the support that the content community provides at conferences like this provides a feel-good boost that lasts for months afterward.

Create a Community of Practice Around Content

Content research works best when you have a community of people supporting each other. You can easily create a Community of Practice around content research in your organization; it's up to you how casual or formal you want or need it to be. However you structure it, make sure that it's providing the moral support and tactical support for those doing the work of content research, and that it provides positive inspiration for brand-new or novice content researchers.

You can make your Community of Practice as simple as a dedicated email alias or Microsoft Teams or other communication channel that's moderated by an experienced researcher. It could be as formal as a regular, monthly meeting where you share best practices, tips, and results with others across your entire company. It could be in the middle—a quarterly or every-other-month check-in with your immediate content team to make sure you're tracking toward the goals set on your content testing roadmap.

The Community of Practice doesn't need to be restricted to just content folks. Product managers, project managers, visual designers, user researchers, and software developers, among others, may be interested in getting involved. (Send an invitation to your CEO and see what happens!)

Some other ideas for making your Community of Practice fresh and interesting:

- **Guest stars:** Invite people from outside your company to share their success stories and hard-won knowledge around the practice of content research.

- **Evolve your techniques:** Focus on one or two content research approaches and ask colleagues to share their specific experiences with those techniques.

- **Platform updates:** Tools like Qualtrics and UserZoom continually evolve their capabilities and features. (Sign up on their website for regular updates and news; they frequently offer free webinars, case studies, and information about new feature releases.)

- **Book club:** Have the team read specific chapters from a content design or content-strategy focused book. (*Writing Is Designing* by Michael Metts and Andy Welfle is an excellent title for content professionals that's from the same publisher as this book.) Take specific tips and suggestions from the book everyone reads and incorporate them into your content research practice.

- **Webinars and workshops:** The UX Content Collective (www.UXContent.com) and other usability-focused workshops are great resources for content professionals. Seattle's School of Visual Concepts (svcseattle.com) offers content and design workshops and certification programs, including a content design track.

- **Meetup share-out:** Meetup.com offers hundreds of monthly special-interest sessions, most of which are free or that charge a small fee. (One challenge with meetups is they're often held after dinner hours, which doesn't work for colleagues with young children or other responsibilities.) When someone on your team can attend a meetup, ask them to share a summary of the topic, what they learned, and how that information can be woven into your team's content research practice.

- **Content research fair:** Create a virtual or hybrid event focused on content research, aimed at helping stakeholders and colleagues outside the practice of content development to understand the power and impact that content research can have on the product and customer experience.

- **Impact round-up:** You know it's important to regularly share details with your broader business group or team about the impact of your content research. When you have a large content team with people assigned to disparate products or business groups, it's easy to lose sight of the collective and awesome impact of content research.

In an impact round-up, you gather slides or details from *all* the research done over a specific period of time—a month, a quarter or two, or even a year—and summarize it in one amazing, energizing, inspiring presentation.

Schedule a meeting and invite any and every person in your organization whose role is affected by content. (Hmm, if you think about it, that might be every single person, right?)

It's fun to have each person who led a research study quickly present their own work: A "Before" view of what content looked like prior to research; an "After" shot of the transformation brought about by the content research; and finally, a quick summary of how the content improvements led to business impact—whether it's increased numbers of customers, increased customer engagement stats, increased revenue, or overall revenue boosting. As an added zing, you can calculate the total impact from every content research study and present that as your parting slide. Ka-pow!

Making It All Work for You

However you approach it, building and keeping up momentum for your content research practice will take dedicated effort. But what you'll find, over time, is an undeniable surge of respect for your content team, which will uplift you and your content coworkers, providing energy and moral support.

Content research helps make the painstaking, complex work of digital content development visible to those outside your content team. It's a direct way to showcase the power of content to create stronger customer experiences and drive business success—and that's a recipe for bringing more influence, power, and success to you and your content team.

APPENDIX

A Little Something Extra

Content Research Terms and Definitions

A/B or multivariate experimentation: A/B experimentation is a rigorous, time- and resource-intensive way of evaluating how engaging or not your content is to your audience or customers. In A/B experimentation, a percentage of customers will be shown Content Version A and another percentage will be shown Content Version B. When many versions of content are included in a single experiment, it's known as *multivariate experimentation*. While A/B experimentation can be valuable, it can be risky because you're showing customers an unproven version of content. (Your audience may also be confused if they view a temporary, experimental version of content on one day, and return to your site and then can't find the content they were expecting.) The risks of A/B experimentation can be minimized by first using content research to identify and "pre-test" content to determine which versions, if any, are most engaging to and effective with your specific audience.

Accessibility: Accessibility in the context of UX content refers to how well content can be used by all people—in particular, those who use assistive technology, such as Braille readers or screen readers. WCAG (Worldwide Content Accessibility Guidelines, pronounced "wuh-kag"), created by the World Wide Web Consortium, exists to standardize and promote best practices for digital accessibility, including those specific to content.

Actionability: Actionability refers to how clear the next steps are for customers to take after reading a specific piece of content. For example, is there a link or call-to-action button for them to select so that they can easily keep progressing through the experience you (the content creator) want them to? Actionability also includes whether your user has enough context and information to take the next step. If users are feeling confused or overwhelmed, they may first go searching for additional information, or may simply give up and abandon your website or app.

Call to action (CTA) or calls to action (CTAs): Typically, a CTA button is used to help your audience "click through" or progress to the next step or screen you want them to view. Sometimes, the process of content creation is so complex that a CTA can easily be inadvertently omitted. It's also easy to have too many CTA buttons or links, which can confuse or overwhelm your audience.

Clarity: Clarity reflects how easy or not it is for customers to understand your content. Several tools exist to help measure clarity, from five-second tests to the Hemingway app to the Flesch-Kincaid formula to SUS (System Usability Scale) and SUPR-Q (Standardized User Experience Percentile Rank Questionnaire).

Cloze tests: Cloze tests are used to evaluate the readability of content. For a cloze test, you take a portion of the content you want to evaluate (usually 250 to 300 words or so) and replace every fifth or nth word with a blank space. Ask participants to fill in those blank spaces. To score a cloze test, take the number of correct responses and divide that by the total number of blank spaces. Scores of 60% or higher typically indicate a good amount of comprehension, while under 40% indicates the text is likely too difficult for the audience.

Conciseness: Conciseness (sometimes called *concision*) refers to how brief content is (while still being clear and complete). Content clarity is more important for the customer experience than content conciseness.

Content asset: A content asset is an individual piece of content at the content-format level, such as a landing page, email, webinar, infographic, e-book, video, podcast, and so on. When creating content, content designers and content strategists consider what ideal format a content asset ought to take, based on the customers' needs. For example, videos are excellent for helping explain step-by-step tasks.

Content audit: A content audit involves evaluating a specific body of content you and your team are responsible for. An audit can be comprehensive and involve your entire content customer experience or your whole app or website, or it may be what's called a *spot audit*—focused on one scoped, specific portion of content. It can include evaluating content for quality (based on your content principles or heuristics), whether it's stronger or weaker compared to your competition, and whether the body of content is lopsided, based on content type (for example, if 60% of your content is webinars, that is likely worth investigating). Often, content audits reveal strengths and areas of opportunity with your content platforms. If it's quite easy for your team to create webinars or blog posts, it's not surprising that you may see a preponderance of that content type revealed by an audit. For more guidance on content audits, see Paula Ladenburg Land's excellent book, *Content Audits and Inventories: A Handbook.*

Content Community of Practice: A content Community of Practice (*CoP* for short) is a morale-boosting support group. Many content Communities of Practice focus on how to raise the profile of the practice of content. If your company is small, you may want to create a cross-company content Community of Practice that welcomes all sorts of content creators: user experience, help content, technical documentation, social media and external communications, internal communications, marketing, and more. If your company is larger, it may make sense to have several coexisting communities of practice, each with their own specialty.

Whatever your company size, make sure to cast a wide invitation and make the Community of Practice a welcoming, inclusive group. Members of a Community of Practice meet regularly to share challenges and suggest solutions. Some also invite guest speakers from outside companies and organizations, to learn about what tools, practices, and processes are working for them.

Content design: Content design is the art and science of determining what specific content should appear in a user experience, and where and when it should appear. Content designers consider how much information and detail a customer needs, based on the knowledge the customer has and the context of the customer experience.

Content heuristics or principles: Content heuristics or principles are the core qualities and best practices you and your content team identify as being central to creating an effective customer content experience. Content heuristics can include many elements, among them content accessibility, accuracy, actionability, alignment to brand guidelines, clarity, completeness, conciseness, customer-centricity, empathy, inclusiveness, relevance, and usefulness.

Content research or content testing: Content research is the practice of evaluating your content with the participation of your audience or customers (or people who are as similar to your audience or customers as possible) with the intent of understanding how your audience is reacting to your content and why. Combining quantitative questions with qualitative questions is a best practice.

Five-second test: This is a type of content research in which a content asset is shown to research participants for just five seconds. Subsequently, participants are asked what, if anything, they can recall about the content, especially if anything made a positive or negative impression on them. Five-second tests are particularly effective for determining whether core messaging of home pages, app start screens, or landing pages is clear and effective or not.

Gaze plot: Gaze plots are a type of quantitative measurement that show where there may be "hot spots" or specific areas of focus on your app or website that are capturing users' attention. For example, a gaze plot may show that a large percentage of your audience is looking at a specific word on your home page or app. (Often, call-to-action buttons are "hot spots" on a gaze plot.) Gaze plots can reflect positive or negative attention, so they may require follow-up questions to understand your users' state of mind. For example, users may be focused on a specific navigation label. That focus may be a signal of interest and engagement or may indicate that the label is unclear to your audience.

Heat map: A heat map is a type of quantitative measurement that shows where your audience or customers are interacting with your content—whether hovering their mouse or finger over a particular section of a page or screen, or where they are tapping or clicking through on calls to action.

Most Important Content (MIC): The MIC is the content that's most essential to your business success. It may or may not include your home page or app start screen. It may or may not be your most highly trafficked or visited content, especially if you have search engine optimization (SEO) challenges. You and your team need to do deep thinking to identify your MIC and prioritize its optimization.

Net Promoter Score (NPS): Net Promoter Score is a measurement of how likely your customers are to recommend your product or service to a friend or family member. It's a customer experience measurement that's frequently used by businesses to evaluate user experiences, with mixed results. Research has shown that when user experiences are objectively improved, NPS can drop instead of increase.

Qualitative research: Qualitative research helps you understand why your audience thinks or acts the way they do. Qualitative research is known as the "why" research, that can embellish or add color to quantitative research, which measures the "what," meaning what percentage of your audience thinks or acts a certain way. The output of qualitative research is often direct quotes, or verbatim comments, from your audience. It's important to note that qualitative research doesn't have anything to do with quality; it is, however, often qualitative research that will provide golden nuggets of insight that can dramatically improve your content experience.

Quantitative research: Quantitative research evaluates how many or what percentage of your audience thinks about or reacts to your content in a specific way. It's also known as the type of research that answers "what" types of questions, compared to qualitative research's "why." The output of quantitative research is often numbers: what quantity or percentage of your audience thinks or acts a specific way.

Scale questions: A scale question (also known as a *Likert scale question*) is a type of quantitative research question that helps you understand levels of degree. They typically include a 5, 7, or 9-point scale, to easily measure participants' negative or positive reactions. An example of a scale question is: "On a scale of 1 to 7, with 1 being not at all likely and 7 being extremely likely, how likely or not would you say you are to select the 'Buy now' button on this website?" *Note:* The words scale, rating scale, and Likert scale can all be used interchangeably.

Screener question: A screener question is a question asked of UX research study participants, to ensure that they meet the criteria needed or wanted by the person(s) running the study. For example, if you are running a study and want only residents of The Bronx, Brooklyn, Queens, and Staten Island to participate, but not residents of Manhattan, you can ask participants which borough they live in. Then you (or your online testing platform) accept only those people who live in the boroughs you want into the study, and do not accept those people who live in Manhattan.

Search engine optimization (SEO): SEO is the art and science of helping your content be found easily and quickly through a web search. There is no shortage of "thought leadership" around SEO. Some of the more trustworthy sources of best practices in this ever-evolving and changing field are Google's Search Central (https:developers.google.com/search), Moz, and Search Engine Journal.

Sentiment research: Sentiment research evaluates how individuals react emotionally to your content. It is sometimes referred to as *hedonic research*. A framework called Plutchik's Wheel of Emotions is often used in sentiment research, as it lists a variety of emotions on a scale from less intense to very intense and can be used as a prompt to encourage participants to identify specifically how they're feeling.

Standardized User Experience Percentile Rank Questionnaire (SUPR-Q): SUPR-Q is a set of eight standard scale questions, developed by John Sauro, that measures usability, trust or credibility, and whether or not customers are likely to be loyal customers, based on their responses to the questions. Like SUS, SUPR-Q can be baselined and remeasured periodically (usually quarterly) to determine whether the customer experience is improving over time.

System Ease Score (SES): System Ease Score is a rating-scale measurement of usability. Typically involving a 5- or 7-point scale, it asks individuals to rate how difficult or easy a single task from a specific user experience was for them. Like most quantitative measurements, its usefulness is amplified when it's combined with an open-ended, qualitative question that asks participants to provide additional detail about why they answered the question the way they did.

System Usability Scale (SUS): System Usability Scale is a standardized set of 10 scale-based or Likert questions, created by John Brooke in 1986. You can ask your customers the SUS questions, as they refer to a specific content experience or piece of content that you want to evaluate. SUS helps you understand how useful and clear the content is for your customers, and it can be useful to benchmark and re-evaluate periodically to gauge whether or not your content is continually improving, according to your customers' points of view.

Verbatims: Verbatims are direct quotations gleaned from qualitative research (open-ended questions). Verbatims help you understand more thoroughly what resonates with users, what confuses them, and perhaps most importantly, your users' emotional state of mind while they're reading your content.

A Quick Guide to Content Research Types

What this guide will help you learn:

- When to run content research and when *not* to
- When to involve your user research team
- The basic research/test-question types, when to use each, and what to expect as outputs (data visualization types)

What is content research and testing?

- Content research and testing is the process of asking your customers or audience—or reasonable proxies for these people—how well or *not* your UX content is resonating with them.
- It's a great way to be customer-centric in your user experience and content design work.
- One of its many benefits is that it will challenge your biases and that of your stakeholders.
- It can be done in many ways, including using online platforms like UserZoom, dscout, and UserTesting.

When should I use content research and testing?

- Content research is helpful to use anytime you don't know for sure how a word or phrase is perceived by your audience (in other words, often).
- Content research is also helpful for any major launches, such as new products, feature releases, or critical business campaigns.

When should I involve my user researcher, instead of tackling content testing myself?

- You should involve your user research lead whenever you are creating a content research/test plan. Ask them to review your plan to help reduce your own bias and to get their feedback on your testing approach. You can run the research study, analyze the results, and write up the report on your own. However, it's always a good idea to get a gut check and feedback on your test plan from the user research team prior to starting any study, if you have the time and they're available to do so.

It's also a good idea to get input from your core product manager, product designer, and lead software developer, if possible. That way, the full feature team is aware that testing is happening, so nobody feels left out or surprised when results are shared.

Your user researcher should also take the lead on testing in these cases:

- If the research study is complex enough to require a moderator (for example, needing a user researcher to guide each participant through the steps involved).

- You're looking for input on *navigation placement:* Determining *where* to place a left-navigation element is quite complex and is best left to the user researchers. (However, if you are curious *what to name* a left-navigation label, content research can be very helpful.)

Seven Types of Content Studies and When to Use Each

These are the seven core types of content research question types:

1. Clarity
2. Comprehension
3. Naming
4. Preference
5. Audience-Specific
6. Sentiment (also called *hedonic* research)
7. Actionability

Clarity Studies

When to use a Clarity test: When you want or need to know whether your users clearly understand a certain word or phrase.

Clarity tests are specifically good for:

- New product or feature names
- Left-navigation labels
- Other areas where understanding the specific word or phrase in question is essential to customer success (just about anywhere)

A. Multiple-choice with image:

Take your design file and scrub out where the specific word or phrase in question will appear. Highlight the spot with a colored box or otherwise mark it in an obvious way (ex: "AAAAA" or "BBBBB").

Include this image file in the research study setup steps.

Use a *multiple-choice* or *written response* question to ask participants what word they would choose to put in the marked spot.

- If you use a *multiple-choice* question, it's helpful to use an open-ended *written-response* question that asks participants *why* they chose the word they did. ("Tell us a bit about why you chose the response you did. Provide as much detail as you'd like.")
- You can also ask another follow-up *written response* question, asking participants to use the word/phrase they chose in the multiple-choice question in a sentence. This will clarify whether they clearly understand its meaning, or if they need to hedge or are only partially able to define it in their own words.

Multiple-choice:
Donut graph
Written response:
Verbatims (direct quotes)
Scale: Bar graph

B. Multiple-choice (no image)

Follow instructions for Part A without using an image.

Use a *multiple-choice* question format. Be sure to provide sufficient context so participants are clear about what the scenario is.

C. Written response

Follow Part A with an image. Using a *written-response* question, ask participants what word they would choose for the designated box or spot.

You can follow up with another *written response* question asking why they chose the word they did.

D. Rating-scale question

Use a *scale* question format. Example: On a scale of 1 to 7, with 1 being not at all clear and 7 being fully clear, how clear do you feel "[the word or phrase that you're testing]" is?

Directly follow up this quantitative question with a *written-response* question asking the participant why they answered the way they did, and ask if they have suggestions for alternate words/phrases. Encourage them to provide as much detail as they would like.

Comprehension Studies

When to use a Comprehension test: When you want to find out how deeply your audience is understanding a word or phrase. Can customers or people in your target audience thoroughly explain the word, phrase, or concept in their own words or use it in a sentence to show they're quite clear on its meaning? This is also known as the "prove it to me!" test format.

Comprehension tests are also helpful for key Jobs to Be Done research. For example, if you are creating content for an online real-estate website and want to know if customers understand the differences between the principal and interest in their monthly mortgage bill, you can ask them to describe these two terms in their own words.

Comprehension tests are specifically good for:

- Table or chart-column labels (Example: For invoices, when testing column labels on the invoice's table, ask users to calculate a subtotal or total.)

- More-complex interactions (Example: In a licensing-assignment scenario, ask users to calculate the total number of users who would have a specific license type assigned to them.)

Which Question Format to Use for Comprehension Tests	Outputs
A. Basic Comprehension tests: Using a *written-response* question format, ask participants to explain a word, phrase, or concept to you to show that they clearly understand their meanings. For more complex scenarios, you can use the *multiple-choice* question format to provide suggested definitions. (Be careful of leading questions. Asking for written responses instead of using the multiple-choice question format can often lead to more easily actionable insights.)	*Multiple-choice:* Donut graph *Written response:* Verbatims (in an Excel spreadsheet)

continues

B. Five-second tests	Verbatims of what specifics (if any) were noticed by participants.
Use the five-*second test* format (if available in your online research platform) and an image of your content. Make sure that the image is zoomed in enough to view clearly.	
Participants will be shown the image for only 5 seconds.	
Then ask participants to describe what they saw and what they recall (if anything) from the content they viewed.	
To get a feel for how the voice and tone of the content is being perceived by participants, you can also ask them how they would describe the overall "feel" of the content: What 2-3 adjectives describe the content from their point of view?	
C. Cloze tests for comprehension	Verbatims of which words participants chose to fill in the blanks.
Using a cloze test format, omit every 5th or nth word. Ask participants to fill in the blanks with the words they think are most clear and suitable.	Overall cloze test score.
To score a cloze test, take the number of correct responses and divide that by the total number of blank spaces. Scores of 60% or higher typically indicate a good amount of comprehension, while under 40% indicates the text is likely too difficult for the audience.	

Naming Studies

When to use a Naming test: They're helpful when you want to know what word to use for a UX element that's especially important to customer success.

Naming tests are specifically good for:

- Left-navigation labels
- Chart or table-column labels
- Features (but be sure to partner with your core UX stakeholder team when doing so).

Which Question Formats to Use for Naming Tests	Outputs

A. Naming test (with or without image)

Use a *multiple-choice, written-response,* or *scale* question format. Using the same techniques as for Clarity tests above (A–D), ask participants what they would choose to name a particular feature, navigation label, etc.

Sometimes, Naming studies require an image or screen grab for reference.

Multiple-choice: Donut graph

Written response: Verbatims (typically in an Excel spreadsheet)

Scale: Bar graph

Preference Studies

When to use a Preference test: When you want to know which specific word or phrases your users say they tend to prefer. Preference tests can be powerful to run instead of A/B experiments. They can also be run prior to A/B or multivariate experiments, so that you can identify words that resonate with your audience and use those words in the A/B experiment to run against the "control" version of content.

Preference tests are specifically good for: Pinpointing what words will enable customer success in essential workflows or Jobs to Be Done.

In Chapter 1, "The Power of Content Research," the study comparing "seat" and "license" is an example of preference research.

Which Question Format to Use for Preference Tests	Outputs

A. Preference test (with or without image)

Provide a screen grab of your design file, with the word or phrase in question scrubbed out, and highlight or otherwise mark this specific spot in the file. (A blank box can work.)

Using the *multiple-choice* question template, ask research participants which word they prefer for that spot.

You may include a "none of the above" option.

Follow up the initial question with a *written-response* question to ask participants why they chose the response they did. If they replied, "none of the above," ask them to suggest the word or phrase they feel is best, if they can think of one.

You can also use an open-ended question format instead of the multiple-choice format. To do so, ask participants what word or phrase they would suggest for the word or term in question.

Multiple-choice: Donut chart

Written response: Participant verbatims (in an Excel spreadsheet)

Audience-Specific Studies

When to run an Audience-Specific test: When you need to confirm how a specific targeted audience responds to your content. For example, if your company creates products for businesses of various sizes—very small businesses, medium-size businesses, large ones, enterprise-size businesses—and are curious if you need to personalize or customize your content for each audience for it to be understandable, clear, and effective.

Another example is if you have brand-new customers using your product for the first time and experienced customers who've used your product for many years. Can you use the same language and word choices for different audiences, or do newer customers require additional explanations, definitions, and context?

Audience-Specific tests are specifically good for:

- Running Clarity, Preference, Comprehension, or Naming tests for a specific audience, whether the specificity varies by business role, company size, or industry.

- Audience-Specific tests can also be run to validate whether a word or phrase resonates with or is clear and understandable to people living in specific geographies (to validate or do quality assurance checks on the work of your localization or translation teams).

Which Question Formats to use for Audience-Specific Tests	Outputs
A. Audience-Specific or targeted test: You will likely need to use screener questions for Audience-Specific studies. Use the audience targeting criteria in your online platform and customize it as necessary using screener questions to accurately target your test to the appropriate audience. Examples: (Note that the targeting of Audience-Specific tests will vary depending on how your company defines your various audience segments.) For very small businesses, target your study to participants in companies of 1–9 employees. For small and medium businesses, target participants in companies of 10–199 employees. Audience-Specific tests can also be used to uncover insights from people who work in specific roles. For example, do C-suite employees have differing perceptions of specific words or phrases than middle managers or individual contributors?	Same as Clarity, Comprehension, Naming, or Preference tests

Sentiment or Hedonic Studies

When to use a Sentiment test: When it's helpful to understand how content makes a user feel. Do specific words or phrases lead users to feel confident, pleased, wary, happy, unhappy, etc.? Often, a framework called *Plutchik's Wheel of Emotions*, which lists emotions like apprehension, serenity, interest, and annoyance on scales from less intense to more intense, is used to help encourage participants to easily identify their state of mind or emotions.

Sentiment tests are specifically good for: Naming or Preference tests. They can help understand the *why* behind user preferences.

Which Question Format to Use for Sentiment Tests	Outputs
A. Sentiment or Hedonic tests Using a *written response*, *multiple-choice*, or *scale* question, ask participants how they feel after reading or interacting with your content. *Caution:* This type of test may be *leading*, meaning that it puts words in the mouths of participants. It is sometimes best to run Sentiment tests *after* running and evaluating any Clarity, Naming, or Preference tests. You can also run a Sentiment test by sharing a content asset with participants and then use a *written-response* question that asks them to list out the words in the content, if any, that elicit certain emotions or strong emotions.	*See above*
B. Highlighter tests Provide participants with a copy of your content (paper or digital). Ask them to use *highlighters* to indicate words or phrases that elicit a particular response. Use one highlighter color per emotion. Example: Participants use a green highlighter to mark the words that make them feel confident, and a yellow highlighter to mark the words that make them less confident or confused.	*This type of test requires compiling/collating the results on your own. You can code the results by making a spreadsheet to note which words elicited which type of results in X number of participants.*

Actionability Studies

When to use an Actionability test: When you want to measure how likely participants say they are to take a specific action.

- Example 1: Finding out whether a user is likely to purchase a product in Purchase services.
- Example 2: Finding out whether a user is likely to read Help content (or right-panel content, or any other element in a specific workflow or scenario).
- Example 3: Finding out what action a user will take after viewing content in a table.

Actionability tests are specifically good for: Determining whether users are able to discover and understand UX content elements that will help them take action. That next action may be selecting a call-to-action button (based on its text) or simply reading content (such as from a fly-out).

- Example 1: A "snowman" within a table
- Example 2: A caret or chevron that opens a right-side panel

Which Question Format to Use for Actionability Tests	Output
A. Likelihood-to-take-action test: Provide participants with a scenario and any specific context they'll need to clearly understand the situation. You can also share a UX image or screen grab if needed. Using a *scale* question, ask participants how likely they are (or not) to click or tap on a call-to-action button or link, or take another action indicated in the content.	Bar graph
B. Next-action test: Provide participants with a scenario and any specific context they'll need to clearly understand the situation. You can also share a UX image or screen grab if needed. Then, using a *written response* or *multiple-choice* question, ask participants which action they're likely to take next (if any).	*Written response:* Verbatim responses (spreadsheet) *Multiple-choice:* Donut graph

Content Test Template: Product XYZ Naming Study

Type of Content Study: [Example: Unmoderated, 10 participants from very small businesses]

Tool: [Example: UserZoom.com]

Stakeholders: [Names and email—include product management, user research, product design]

Goals

1. What business/customer questions are you trying to answer?
2. What do you hope to learn?

Non-Goals

1. What questions are you *not* answering about this experience (for example, related questions that have already been answered in previous studies)?

Scenario

[Example: You have a small business with just you and 5 employees, and you just signed up to use Product XYZ for the first time. After purchasing it, you are taken to this webpage.]

Tasks for the Content Research Study

Participant Instructions:

1. Click the URL to go to the prototype (or view the screenshot).
2. [Add Prototype URL here—for example, link to Figma screen]
3. Please answer these questions about the information shown:

    [Examples:]

    a. What words would you use to describe this page? (Open-ended question format)
    b. When you look at this page, how does it make you feel about managing Product XYZ for people in your company? (Scale/Likert question, ranging from very anxious [1] to very confident [9].)
    c. Tell us why you chose this answer. If you chose "Other," what would you say instead and why? Walk us through your thinking.

d. Which of these statements seems most appropriate to describe what is shown?

- It's a dashboard.
- It's a home page.
- It's a specific view of the home page.
- It's a mode for viewing the admin center.
- Other.

Audience Targeting

Summary of who the target audience is for the test:

1. Company size (very small, small, medium, large, enterprise)
2. Countries (typically tests are sent to: U.S., Canada, Australia, UK)
3. Income range
4. Employment status
5. Other screener questions to help ensure that the test is sent to people in the audience needed

Screening Questions to Refine Your Audience Targeting

Ex: To target users of Product XYZ, you can ask:

Are you a regular user of the Product XYZ?

Yes (accept)

No (reject)

Ex: To target very small businesses, you can ask:

How many people, including yourself, work for your company?

1–9 (accept)

10–99 (reject)

100–399 (reject)

400+ (reject)

INDEX

feature-focused content research, 105

Federal Trade Commission (FTC), advertising and marketing laws, 13, 46

Fenton, Nicole, 50

file-naming, standardization of, 203

findable content

as content principle, 64–65

and search engine optimization, 63, 90–91

five-second tests, 220, 265, 272

Flesch-Kincaid's formula for reading levels, 35–38

in evaluating content, 76

Grade Levels, 36–38

Reading Ease, 36–38

fluff in content, 29–34, 51

"fly" approach: meta-analysis, 206, 210

focus groups, 133

Forrester Research, on ROI for usability research, 134

friction-filled phrasing, 158

funnel views, 90–92

G

gap analysis, 219

gaze plots, 218–219, 264–265

gender identity in selection of research participants, 143

gerunds (verbs), 40

glossary of terms, 262–267

goals of content research, 5, 97–121

considering altitude, 105–108

eradicating jargon, 109–111

growing your content team, 101

making it all work for you, 121

messaging framework research, 105–107

in outline of test plan, 127–128

product- or feature-focused research, 105

question types, 112–121, 267–269. *See also* research question types

starting small and scoped, 102–104

starting with Most Important Content, 98–102

word- and phrase-specific research, 108

Google Analytics, 88, 89

Google Trends, 82, 86

Google's HEART framework, 88, 93

governance of content, 91, 101, 251

grading content, 75

guerrilla research, 132

guidelines

compared with rules, 34

for creating plain language content. *See* plain language guidelines

Intuit's content design guidelines, 30–31

voice-and-tone. *See* voice-and-tone guidelines

writer's style guides, 13–15, 60, 242–243

H

Halvorson, Kristina, 58–59, 257

hard-to-simplify words, 33

headline, on landing page, 98–99

health insurance example, 2–4, 238

HEART framework, by Google, 88, 93

heat maps, in evaluation of content, 82, 84–86, 216–219, 265

hedonic research, 119–120, 266, 275

help content, as content format in competitive analysis, 225

heuristics, 55–56, 62, 67, 265

highlighter testing, 165, 275

highlighting notable responses, 178–179

longitudinal analysis, 226

Loranger, Hoa, 47

Lostutter, Melvin, 34

low vision, 174

M

macro-level competitive analysis, 222–223

Mailchimp Content Style Guide, 14

Malig, Irish, 178, 190

Manoogian, John, III, 144

marketing laws, 13

marketing team, collaboration with content team, 146

Maslow's Hierarchy of Needs, 64

McBride, Heather, 146

MeasuringU, 228, 229, 230

meetups, 256, 258

Merriam-Webster, 39, 41

messaging framework content research, 105–107

meta-analysis
  for communication of results, 206, 210
  of content research, 226–227

meta description, 99–102

Metts, Michael, 165–166, 258

micro-level competitive analysis, 223

Microsoft invoices, xvi–xx

Microsoft Word, readability statistics, 36

moderated interviews, 232–234

moderated online research, 132–133

momentum building, with content research, 247–260
  create a content research roadmap, 249–251
  create Community of Practice around content, 58–59, 256, 257–259
  document research methods and tools, 248
  include research in product planning, 255–257

making it all work for you, 259–260

train your team, 251–255

Most Important Content (MIC), 74, 98, 221, 228, 242, 249, 251
  defined, 265

Moz, 82, 85

multiple-choice research questions
  analysis of results. *See* analysis of responses from multiple-choice questions
  crafting the questions. *See* research questions, crafting multiple-choice questions
  as preference research questions, 117
  using plain language, 28

multivariate experiments, 23, 116, 172, 244–245, 262

N

naming research, 114–115, 272–273

naming study, 277–278

navigation labels, 52, 113

navigation placement, 269

Net Promoter Score (NPS), 74, 83, 120, 265

newsletter for communication of results, 205, 208

Nicely Said (Fenton and Lee), 50

Nielsen Norman Group
  gaze plot, 219
  "How People Read Online," 34
  plain language preferences, 10, 47
  usability heuristics, 55–56

"noisy" research results, 150

non-content folks (NCFs), 71

nondisclosure agreements (NDA), 137, 233

"none of the above" response option, 151–152

O

online content, reading comprehension compared with print, 34
online dictionary, 39
online research
    moderated and unmoderated, 132–133
    participant pools, 141–144
    platforms, 133–136, 230
open-ended questions
    analysis of results. *See* analysis of responses from open-ended questions
    crafting the questions. *See* research questions, crafting open-ended questions
open house, for communication of results, 213

P

page title, for SEO, 99–100
person-on-the-street interviews, 132, 133
personally identifiable information (PII), 136–137
phrase-specific content research, 108
pie chart, how to create, 174
ping-ponging, 89
Plain English Campaign, 51
plain language, 27–52
    content fluff, 29–34
    eighth-grade reading level, 34
    Flesch-Kincaid's formula, 35–38
    guidelines for creating content. *See* plain language guidelines
    hard-to-simplify words, 33
    legal team and legalese, 45–46
    making it all work for you, 51
    measuring content clarity, 35–38, 76, 80
    Microsoft Word readability statistics, 36
    not dumbing down, 29, 166

in research, multiple-choice questions, 28
resources, 50–51
support for, 165–166
"you are not your audience," 30
plain language guidelines, 38–50
    active verbs, 40
    clear and understandable words, 47
    contractions, 40–41
    jargon, 41–43
    Latin terms, 43–44
    prepositional phrases, 48–50
    short sentences, 47–48
    short words, 38–39
"Plain Language Is for Everyone, Even Experts" (Nielsen Norman Group), 10
Plutchik's Wheel of Emotions, 266, 275
Pomodoro Technique, 201
pools of online research participants, 141–144
post-research tasks, 235–245
    balancing urgency in content update process, 236–238
    informing A/B and multivariate experiments, 244–245. *See also* A/B experimentation
    making it all work for you, 245
    showing the impact of updates, 239–242
    taking credit for your work, 241–242
    updating content guidelines and style guide, 242–243
power of content research. *See* content research, power of
preference research, 116–117, 273
preference testing, 5, 66
prepositional phrases in plain language guidelines, 48–50
principles. *See* content quality principles

product development process
including research in product planning, 255–257
working within to update content, 236–238

product-focused content research, 105

product manager, understanding value of content research, 21

profiles of research participants, 139, 140–141

publishing customer-facing content, working within product development process, 236–238

Q

qualitative data, from open-ended research questions, 155, 161–162, 176

qualitative research
defined, 265
evaluation of, 189
as "why," 7–10, 106

quality principles. *See* content quality principles

Qualtrics, 165, 258

Quantifying the User Experience (Sauro and Lewis), 84

quantitative data
from multiple-choice questions, 168
from scale questions, 175

quantitative information, 5

quantitative research
defined, 266
evaluation of, 188
as "what," 5–6, 9, 10

questions
for content research. *See* research questions
screener questions for participant selection, 140–141, 266
types. *See* research question types

R

RACI chart for research planning, 125–126, 131

racial diversity in selection of research participants, 144

radar graphs, 94–95

rating scales for content evaluation, 75, 77
analysis of responses to scaled questions, 175–176
crafting questions. *See* research questions, crafting rating-scale questions

readability and clarity measurements, for content evaluation, 35–38, 76, 80

readability scales, 35–38

readable, content principle, 65

The Reader's Brain: How Neuroscience Can Make You a Better Writer (Douglas), 50

reading levels
in evaluating content, 76
origins of eighth-grade goal, 34

Redish, Janice (Ginny), xiv-xv, 51

report of research results. *See also* sharing results from studies
consistency of research communications, 199
saving for sharing and for future reference, 202–203
standardized file names, 203
using research plan in writing, 201–202

research participants, 136–144
being respectful of, 139, 165
bias in selection of, 137, 143–144
compensation for, 142–143
current customers as, 137
limiting number of, 139–140
and nondisclosure agreement (NDA), 137

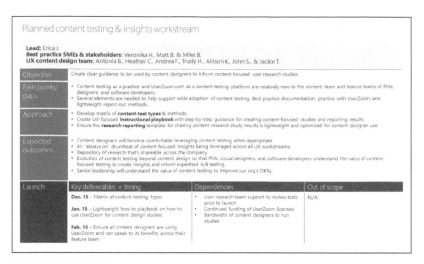

Planned content testing & insights workstream

Lead: Erica J.
Best practice SMEs & stakeholders: Veronika H., Matt B. & Mike B.
UX content design team: Antonia B., Heather C., Andrea F., Trudy H., Allison K., John S., & Jackie T.

Objective	Create clear guidance to be used by content designers to inform content-focused user research studies.
Pain points/ gaps	• Content testing as a practice and UserZoom.com as a content-testing platform are relatively new to the content team and feature teams of PMs, designers, and software developers. • Several elements are needed to help support wide adoption of content testing: Best practice documentation; practice with UserZoom; and lightweight report-out methods.
Approach	• Develop matrix of **content-test types** & methods. • Create UX-focused **Instructional playbook** with step-by-step guidance for creating content-focused studies and reporting results. • Ensure the **research reporting** template for sharing content research study results is lightweight and optimized for content designer use.
Expected outcomes	• Content designers will become comfortable leveraging content testing when appropriate. • An "always on" drumbeat of content-focused insights being leveraged across all UX workstreams. • Repository of research that's shareable across the company. • Evolution of content testing beyond content design so that PMs, visual designers, and software developers understand the value of content-focused testing to create insights and inform expedited A/B testing. • Senior leadership will understand the value of content testing to improve our org's OKRs.

Launch	Key deliverables + timing	Dependencies	Out of scope
	Dec. 15 – Matrix of content-testing types **Jan. 15** – Lightweight how-to playbook on how to use UserZoom for content design studies **Feb. 15** – Ensure all content designers are using UserZoom and can speak to its benefits across their feature team	• User research team support to review tests prior to launch • Continued funding of UserZoom licenses • Bandwidth of content designers to run studies	N/A

Prioritized roadmap

	Step	Priority	Details/Deliverables/SMEs	Timeline	Expected impact
STEP 1	**UserTesting access**	Complete	• Content designers need access to the UserZoom platform. They also need to set up their username and password for UserZoom, which can be complicated.	Oct.	Content designers will have the ability to create new tests and get access to the team repository of past tests & data.
STEP 2	**How-to: Types of content testing**	P0	• The content team needs to know what types of content testing are feasible. • When using UserZoom, they also need to know when to use the various question formats available on that platform. • **DELIVERABLE:** Matrix of test types, best practices, and how to enable testing for each type	Jan.	Content designers will feel confident choosing the right test type to suit their needs and will be able to suggest content tests to feature teams for broader impact.
STEP 3	**How-to: When to use unmoderated tests on UserZoom**	P0	• Content writers need to know when an unmoderated test is appropriate vs. when a different type of testing is better (in partnership with UX research team). • **DELIVERABLE:** Matrix of content research study types & how to run them	Jan.	Content designers will know when to fully engage the user research team for a study vs. when only consultation is needed.
STEP 4	**How-to: How to set up and target a test**	P1	• Clear guidance is needed for new users to understand the many steps involved in creating and targeting a UserZoom study. • **DELIVERABLE:** Guide to creating a test, saved to the content testing Teams channel wiki	Feb.	Content designers will feel confident creating and launching UserZoom tests independently.
	How-to: Analyzing content research metrics	P1	• Improve the customer experience for all core product teams by leveraging quantitative and qualitative insights	Feb.	Content designers will understand how to analyze qualitative and quantitative test results.

FIGURE 12.1

A sample content research roadmap shows dates, goals, and people who are responsible for specific tasks.

NOTE **MIND YOUR CONTENT RESEARCH LIMITS**

With most content research platforms, you'll have a limit on the number of research studies you can run in a quarter or year. Therefore, you may need to track your team's capacity for research studies carefully, and fairly allot the total number of studies across the people on your team. Sometimes, it's quite tricky to do this, especially if one content lead is assigned to a high-priority project that warrants multiple content research studies.

writers' guidelines, 13–15, 60, 242–243.
    See also guidelines
Writing Is Designing (Metts and Welfle), 165, 258
"Writing Small" (Intuit), 30–31
Writing Tools: 55 Essential Strategies for Every Writer (Clark), 50
written-response question format, 155

ACKNOWLEDGMENTS

To My Family:

Enormous thanks to my husband, Brian Williamson, who provided encouragement and constructive feedback and kept the stereo spinning with great music as I wrote. You are a gem.

Thanks to our amazing children, Isobel Williamson and Finian Jorgensen. I love every minute of being your mom. (I'm also thrilled that you're both exceptional writers!)

To the Technical Reviewers:

Special, extra-enthusiastic thanks and appreciation to Bryan Perkins, Heather Campbell, and Dylan Romero for their thoughtful feedback and valuable suggestions for additions and improvements.

To My Microsoft Colleagues:

Dozens of current and former coworkers provided inspiration, shared examples of content research in action, and gave hours of their time reviewing early drafts of the book.

Thank you, thank you to my manager and mentor at Microsoft, Sheila O'Hara, who provided constant encouragement throughout the past two years, and set the stage for us to establish an energizing, inspiring content research practice and clearly demonstrated the business results of the team's content research efforts. Extra thanks to Trudy Hakala for your partnership, indefatigable energy, and unwavering dedication to the customer experience. To the rest of the small but mighty Microsoft 365 content design team—Antonia Blume, Heather Campbell, Andrea Fowler, Allison Klettke, Liz O'Connell, and Jackie Tidwell—thank you for being inspiring, collaborative coworkers, and for your enthusiastic embracing of content research!

Thank you to the crackerjack Windows 365 Cloud PC content design team, Kay Jorgensen and Jaime Ondrusek. Your work helped pave the way for content design to have broader and deeper impact across Microsoft.

Thanks to the amazing content designers across the Microsoft 365 organization: Rose Cooper-Finger, Tany Holzworth, Adesuwa Joseph, Paula Margulies Sion, Louie Mayor, Olivea McCollins, LauraNewsad, Tom Resing, Kirsten Rue, Kaarin Shumate, Christie Vukos-Walker, Trish Winter-Hunt, Beth Woodbury, and Wendy Zucker. Long live the Content Community of Practice!

To my Microsoft colleagues in the Cloud Marketing and Microsoft 365 web teams, thank you for your moral support, creativity, and thoughtfulness: Kelly Anderson, Kyle Brand, Rup Choudhury, Koreen Hirakami, Jenny Hsu, Tiara Jewell, Angela Lean, Charlene Marsh, Tyler Mays-Childers, Jessica Miller, Jackie Ostlie, Jenny Passero, Janine Patrick, Bryan Perkins, Silke Rybicki, Cyndee Settle, and Derek van Veen.

To my Microsoft colleagues across the company—thank you for your enthusiastic support and collaboration, and for your unwavering focus on the customer experience: Shary Almaguer Leal, Holly Ambler, Ryan Baker, Peter Baumgartner, Cynthia Borsheim, Matt Brodsky, Ph.D., Bhavya Chopra, Carrie Cosgrove, Doug Eby, Abena Edugyan, Jonathan Foster, Alex Fromm, Lauren Gallagher, Philip Gerity, Pam Green, Kylie Hansen, Veronika Hanson, Deborah Harrison, Perry Holzknecht, Ivaylo Ivanov, Mike Jeffers, Karen Kesler, Doug Kim, Terry Kirkwood, John Martin, Cern McAtee, Yalda Modarres, Manal Mohanna, Christian Montoya, Joydeep Mukherjee, Kyryl Nagaichouk, Amanda Neves, Kate O'Leary, Eric Orman, Deborah Pisano, Kelley Rand, Tristan Scott, Daniel Simpson, Veronika Sipeeva, Joe Smith, Serdar Soysal, Sindhia Thirumaran, Scot Vidican, Laura Williams, Ph.D., Aaron Woo, and Elaine You.

And thank you to Ana Arriola-Kanada and Alex Lopez for welcoming content design into the Microsoft Design Week programming, and an extra shout-out to Trish Winter-Hunt for your partnership in bringing the content design track to life.

To My Premera Blue Cross Colleagues:

Thank you to these caring, customer-focused, creative, and kind people: Jen Betterley, Kathryn Brookshier, Carleigh Burfitt, Sara Correa Bell, Sarah Coan, Johanna Dokken, Elaine Helm, Todd Johnson, Curtis Kopf, Irish Malig, Molly Marsicek, Sarah McIlwain, Candace Nelson, Hillary Omdal, Laurie Pritchard, Karma Raad, Robert Racadio, Ph.D., Ryan Rosensweig, Melissa Serdy-Velez, Ankit Shah, Veronika Sipeeva, Tom Thompson, and Maxine Williams.

To the Rosenfeld Media Team:

Last, but not at all least, thank you to the crew at Rosenfeld Media for your hard work and unwavering support. I'm so grateful for our partnership on this book. I'm especially thankful for managing editor Marta Justak and her dedicated, thoughtful feedback and continual encouragement throughout the book writing and editing process.

 Rosenfeld®

Dear Reader,

Thanks very much for purchasing this book. There's a story behind it and every product we create at Rosenfeld Media.

Since the early 1990s, I've been a User Experience consultant, conference presenter, workshop instructor, and author. (I'm probably best-known for having cowritten *Information Architecture for the Web and Beyond*.) In each of these roles, I've been frustrated by the missed opportunities to apply UX principles and practices.

I started Rosenfeld Media in 2005 with the goal of publishing books whose design and development showed that a publisher could practice what it preached. Since then, we've expanded into producing industry-leading conferences and workshops. In all cases, UX has helped us create better, more successful products—just as you would expect. From employing user research to drive the design of our books and conference programs, to working closely with our conference speakers on their talks, to caring deeply about customer service, we practice what we preach every day.

Please visit rosenfeldmedia.com to learn more about our **conferences**, **workshops**, **free communities**, and **other great resources** that we've made for you. And send your ideas, suggestions, and concerns my way: louis@rosenfeldmedia.com

I'd love to hear from you, and I hope you enjoy the book!

Lou Rosenfeld,
Publisher

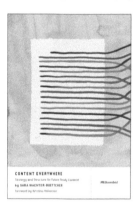

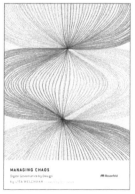

ABOUT THE AUTHOR

Erica Jorgensen lives in Seattle with her family. Her entire career has been focused on working with words—first as a journalist and then in content strategy and UX content at companies such as Amazon and Expedia. She spent five years at Microsoft, where she created user experiences used by millions of customers and expanded the profile and impact of the content design practice. Erica speaks frequently at content and experience design conferences like Button and UXDX and also facilitates content workshops to companies of all sizes. She lives in Seattle, Washington, and can be reached at EricaJorgensen.com.

CPSIA information can be obtained
at www.ICGtesting.com
Printed in the USA
JSHW041158100723
44225JS00001B/1

9 781959 029571